Molecular Orbital Calculations for Amino Acids and Peptides

Anne-Marie Sapse

Molecular Orbital Calculations for Amino Acids and Peptides

With 32 Figures

Birkhäuser
Boston • Basel • Berlin

Anne-Marie Sapse
John Jay College and Graduate School
City University of New York
New York, NY 10019
    and
Rockefeller University
New York, NY 10021
USA

Library of Congress Cataloging-in-Publication Data
Sapse, Anne-Marie.
    Molecular orbital calculations for amino acids and peptides / Anne-Marie Sapse.
        p. cm.
    Includes bibliographical references and index.
    ISBN 0-8176-3893-8 (hardcover: alk. paper)
    1. Amino acids. 2. Peptides. 3. Molecular orbitals. I. Title.
QD431.S257   1999
547′.750448—dc21   99-26375
                                                                              CIP

Printed on acid-free paper.

© 2000 Birkhäuser Boston   *Birkhäuser* 🦉 ®

ISBN 0-8176-3893-8
ISBN 3-7643-3893-8   SPIN 19901572

Typeset by Best-set Typesetter Ltd., Hong Kong.
Printed and bound by Sheridan Books, Inc., Ann Arbor, MI.
Printed in the United States of America.

9 8 7 6 5 4 3 2 1

*To my husband, Marcel Sapse, and to my daughter, Danielle Sapse,
without whose support I could not have written this book.*

Contents

Preface

This book is intended mainly for biochemists who would like to augment experimental research in the domain of amino acids and small peptides with theoretical calculations at the ab initio level.

The book does not require a profound knowledge of mathematics and quantum chemistry. It teaches one rather how to use computer software such as the Gaussian programs and gives examples of problems treated in this manner.

Chapter 1 describes the calculations and one of the programs used for ab initio work.

Chapter 2 describes calculations on small amino acids, such as glycine and alanine.

Chapter 3 discusses the biochemical properties of GABA (gamma amino butyric acid), which is one of the most important amino acids of the nervous system. Ab initio calculations performed in order to study the structure of GABA are presented.

Chapter 4 discusses an amino acid related to GABA, namely DABA (diaminobutyric acid), presenting information about its structure and transport properties.

A number of amino acids, essentials in the biochemistry of organisms, are discussed in Chapter 5. These acids have been subjected to ab initio investigation. Proline, a special amino acid as far as structure is concerned, is discussed in Chapter 6.

Chapter 7 discusses two sulfur-containing amino acids, taurine and hypotaurine, presenting some experimental studies on their mode of action and an ab initio study of their structure.

Starting with Chapter 8, small peptides of great importance are discussed. Glucagon, a small peptide that plays a role in diabetes, is the subject of Chapter 8.

Chapter 9 discusses the pheromone alpha factor, from an experimental and theoretical point of view.

Chapter 10 presents calculations on tight turns in proteins.

Chapter 11 discusses some small peptides that have been studied with ab initio methods.

Oligopeptides that feature anticancer activity, such as lexitropsins, are discussed in Chapter 12.

The book is addressed to graduate and postgraduate students as well as other researchers in the amino acid and peptide area.

New York, NY   Anne-Marie Sapse

Introduction

Knowledge about the origin of life requires the recapitulation of the steps of archaic molecular evolution. According to the protenoid model, proteinoids (copolyaminoacids) arose on earth from mixtures of self-sequencing amino acids. The structure of amino acids, of the peptides formed by their polymerization via the formation of peptidic bonds, as well as the structure of the proteins that are polypeptide chains in various numbers and conformations, have formed the subject of an enormous number of experimental and theoretical studies.

At present, both theoretical and experimental methods are taken seriously as useful sources of information. They compare results and confirm or dispute structural findings. While experimental results are usually not doubted, and computational results depend on such parameters as the quality of the basis sets used, there have been instances in which computational results have contradicted experimental ones regarding structural determination. However, in most instances the two types of methods complement each other. For instance, a laboratory search for intermediates in certain reactions can be avoided once large basis-set calculations show the intermediates not to be a stationary state, more exactly, a minimum on the energy hypersurface.

The application of computational methods to biological systems dates from the 1950s, when the pioneering work of Bernard and Alberte Pullman was first published. The biological systems studied with the quantum-chemical methods available at that time had to be small, and not all the conclusions derived were correct. However, this work opened the door to a whole new area of research.

The basic problem in the determination of the structure of biological systems is their size. In order to be able to handle such molecules as the nucleic acids or the proteins, new theoretical methods had to be developed, and the quantum-chemical methods, ab initio and semi-empirical, were augmented by the molecular mechanics method, which uses experimental parameters in order to determine the force fields of the systems.

Huge strides have been made in the development of computer programs that handle larger systems. Researchers are striving to find the optimum combination of accuracy and expediency, with the ultimate goal being the reduction of computational effort with no loss of accuracy.

All three of these types of theoretical methods are used in the description of amino acids and peptides. The size of proteins precludes the use of ab initio or semiempirical methods, so they are mainly described with computer modeling, with programs such as Sybil, Quanta, and Insight, augmented by energy calculations with the Charmm program and other molecular-mechanic calculations.

The primary structure of proteins, characterized by the amino acid composition and sequence, is determined experimentally by degradation via hydrolysis of the peptidic bonds. The classic method of determining the sequence involves Edman degradation, which is an end-labeling procedure. Physical methods used include mass spectrometry and nuclear magnetic resonance (NMR). Since the 1980s, sequencing of proteins has been performed by sequencing its mRNA or gene.

The three-dimensional structures of about 800 proteins have been determined by Max Perutz and John Kendrew using X-ray crystallography. Recently, NMR methods have also been used. The secondary structure of proteins, with 60% alpha helices or beta sheets and the rest random coils and turns, is determined by the propensity of the amino acids constituting the given protein to form either alpha helices or beta sheets. It is recognized now that the sequence of a protein determines its three-dimensional structure.

Given the size of proteins, quantum-chemical conformational and energy calculations are at present impossible. Some calculations on proteins are being performed at present in Dr. Lothar Schafer's laboratory. Undoubtedly, the increase in computer capacity and progress in computer algorithms will make it possible to perform many such calculations in the not too distant future. The theoretical methods used so far for proteins include molecular-mechanics methods that neglect electrons and describe the motion of nuclei under the influence of an empirical or quantum-mechanically calculated potential energy function, methods that do not use energy functions except in terms of stereochemical principles, computer graphics methods, and molecular-dynamic methods.

Smaller peptides have also been described by the above-mentioned methods, especially the empirical conformational energy program for peptides (ECEPP), written by Sheraga and his group, which has been applied to a large number of small peptides.

In recent years it has become possible to treat amino acids and small peptides with quantum-chemical calculations, as will be described in the next chapters.

1
Theoretical Background

Inadequate descriptions of atoms and molecules by the methods of classical physics led researchers to propose new ways to describe physical reality, giving birth to a totally new science, quantum mechanics. The methods of quantum mechanics are based on the introduction of a wave function, whose physical meaning is related to the probability of finding a certain particle, at a certain time in a volume element, positioned between x and $x + dx$ in the $x =$ direction, between y and $y + dy$ in the $y =$ direction, and between z and $z + dz$ in the $z =$ direction at certain time t. This wave function Ψ satisfies the Schrödinger equation,

$$\left(-\frac{\hbar^2}{2m}\nabla^2 + v\right)\Psi = E\Psi, \quad \hbar = \frac{h}{2\pi},$$

or for short, $H\Psi = E\Psi$, where H, the Hamiltonian operator, is defined by the expression

$$H = -\frac{\hbar^2}{2m}\nabla^2 + V;$$

h is Planck's constant; ∇^2 is the sum of the partial second derivatives with respect to x, y, and z; m is the mass of the particle; and V is the potential energy of the system. The Hamiltonian H represents the quantum equivalent of the sum of the kinetic energy and potential energy, with V being the potential energy operator and $\frac{-\hbar^2}{2m}\nabla^2$ the kinetic energy operator. Finally, E is the total energy of the system and is a number, not an operator.

The wave function, satisfying the Schrödinger equation, and the energy contain all the information about the system within the limits of the Heisenberg uncertainty principle, which states that the exact momentum and position of a particle cannot be known simultaneously. This is why the wave function represents a probability and not a certitude.

Applied to atoms, the Schrödinger equation describes the motion of the electrons in the electrostatic field created by the positive charge of the nucleus. In addition, each electron is subjected to the field created by the negative charge of the other electrons. When the Schrödinger equation is applied to molecules, the motion of the nuclei has also to be taken into consideration, but the fact that the nuclei are so much heavier than the electrons makes it possible to neglect their motion. This is embodied in the Born–Oppenheimer approximation. Accordingly, the electronic distribution in molecules does not depend on the motion of the nuclei, but only on their position. Indeed, the position of the nuclei determines the positive component of the electrostatic field to which electrons are subjected. The kinetic energy operator of the nuclei is considered to be zero.

The many-electron molecule can be thus described by a Hamiltonian written as

$$H = K + V,$$

where K, the kinetic energy operator, is

$$-\frac{\hbar^2}{2m}\sum_i \frac{\partial^2}{\partial x^2} + \frac{\partial^2}{\partial y^2} + \frac{\partial^2}{\partial z^2},$$

with the sum taken over the number of electrons, while V, the potential energy operator, is composed of two electronic terms. One is the attraction between the positive nuclei and the negative electrons, expressed as

$$-\sum_i \sum_I \frac{Z_I e^2}{\mathbf{R}_I - \mathbf{r}_i},$$

where i represents, as before, the summation over the electrons, and I is the summation over the nuclei. Here Z is the charge of the Ith nucleus, and $\mathbf{R}_I - \mathbf{r}_i$ is the distance between the Ith nucleus and the ith electron. The second term represents the repulsion between electrons:

$$\sum_i \sum_{j \neq i} \frac{e^2}{\mathbf{x}_i - \mathbf{x}_j},$$

where $\mathbf{r}_i - \mathbf{r}_j$ represents the distance between electron j and electron j, and e, as before, is the charge of the electron. In addition, one must consider the nuclear repulsion, which determines the nuclear potential energy. This can be expressed as

$$\sum_I \sum_{J \neq I} \frac{Z_I Z_J e^2}{\mathbf{R}_J - \mathbf{R}_I},$$

where $\mathbf{R}_J - \mathbf{R}_I$ is the distance between nucleus I and nucleus J.

The Schrödinger equation can be solved analytically only for one atom: the hydrogen atom. The solution, even for the lightest atom, is complicated, containing spherical harmonic functions and Hermite polynomials. When

the electron–electron interactions are involved, as they must be for any
atom containing more than one electron, the Hamiltonian cannot be
expressed any longer in terms of spherical coordinates, which allow the separation of the three-dimensional form into three one-dimensional solvable
equations. Therefore, a number of approximations have to be introduced.

The main approximation used to solve the Schrödinger equation for
systems larger than the hydrogen atom is the variation principle. Indeed,
when the equation is applied to atoms, the wave function is composed of a
set of functions called atomic orbitals, corresponding to given energy states,
containing a number of electrons determined by Pauli's exclusion principle. If the exact form of Ψ is known, the energy of the system can be computed by using the expression

$$E = \frac{\int \Psi^{\alpha} H \psi d\tau}{\int \psi^{\alpha} \psi d\tau}.$$

If the exact form of Ψ is not known, an educated guess can be taken, and
the approximate value of Ψ is used to compute an approximate E. The variation principle states that the expectation value of the energy thus obtained
will always be higher than the exact energy of the system. This allows the
energy to be minimized with some parameters characterizing the wave
function, in order to obtain the closest possible energy to the exact energy
of the system. This procedure establishes a number of equations whose solutions are the optimum values for the parameters of the system.

One of the methods to construct a good wave function is the
Hartree–Fock method.

The Hartree–Fock method deals with the reason for the impossibility of
solving the Schrödinger equation analytically: the term e/r–r, which is the
term representing interelectronic repulsion. In the absence of this term, the
equation for an atom with n electrons can be separated into n equations
for the hydrogen atom. If the sum of these terms is replaced by the sum of
terms describing the motion of each electron through a cloud of electric
charge due to the other electrons, the equation becomes solvable through
an iterative method. Indeed, the electronic cloud is characterized by its
charge density, which depends on the atomic orbitals describing the electrons. Once the interaction between a given electron and the cloud of the
other electrons is calculated making use of an initial approximated orbital,
the equation can be solved, and a new, improved orbital is obtained. This
new orbital replaces the initial guess in the equation, whose solution represents an even more improved orbital. This iteration procedure is repeated
until a certain threshold is reached.

For molecules, molecular orbitals have to be used instead of atomic
orbitals. These can be built out of atomic orbitals, and one of the most
widely used methods is to construct the molecular orbitals out of a linear
combination of atomic orbitals (LCAO). The total wave function of the

system has to be augmented by spin orbitals called alpha (for spin $\frac{1}{2}$) or beta (for spin $-\frac{1}{2}$).

In order to perform quantum-chemical calculations without using any approximations, such as neglecting integrals of interaction between atomic orbitals located at different centers or using experimental parameters, one has to use the ab initio method, which uses a theoretically constructed wave function from the beginning. Since the Hartree–Fock method involves the calculation of integrals over atomic functions, the computational time is proportional to N^4, where N is the number of atoms of the system. For amino acids, and especially for peptides, this is an enormous task, since the atomic orbitals are exponential functions of the form $e^{-\xi r}$, where r is the distance of each electron from the nuclei. This form, called the Slater orbital, requires a large amount of computer time for the computation of the integrals. To shorten the time, these functions have been replaced by expansions in a certain number of Gaussian functions of the form $e^{-\alpha r^2}$. The integrals over Gaussian functions are much easier to compute. To reproduce better the form of a Slater orbital, which is the real dependence of the functions on r, as large a number of Gaussians as possible has to be used for the expansion.

So the function will take the form

$$\psi = ce^{-\alpha r^2},$$

where α is a constant determining the radial extent and c is another constant.

Among the computer programs devised for performing ab initio calculations are the Gaussian programs, written at Carnegie Mellon University, in Pittsburgh, Pennsylvania. These programs make use of the expansion of Slater-type orbitals into a series of Gaussians, thereby establishing different basis functions for describing the system.

The smallest basis set used by the programs is the STO-3G basis set. The name comes from "Slater-type orbital," expanded into a series of three Gaussians. For hydrogen atoms, the orbital is the s orbital, while for heavier atoms s and p orbitals are used, as appropriate for a given electron in the atom. For large systems, the STO-3G basis set is the only possible one. Slightly larger minimal basis sets include the STO-4G, STO-5G, and STO-6G, where only one Slater orbital is used, expanded into 4, 5, and 6 Gaussians, respectively. It has been found that the energy decreases with the number of Gaussians, but such important information as optimum geometry, energy differences, and atomic charges are fairly insensitive to this number. In most cases bond distances calculated by STO-3G are very close to the experimental ones.

A larger series of basis sets are the split-valence basis sets. Among these, the double-zeta basis sets consist of two Slater-type orbitals for the valence electrons, one expanded in a number of Gaussians, the other approximated by one Gaussian. The core electrons are described by one Slater-type orbital, expanded in a number of Gaussians. For instance, one of the most widely used basis sets, the 6-31G basis set, has the core electrons described by a Slater-type orbital expanded in a series of six Gaussians, while the valence electrons are described by two Slater-type orbitals, one expanded in a series of three Gaussians and the other one approximated by one Gaussian function. The functions used are s for hydrogen and s and p for nonhydrogen atoms. Triple-zeta basis sets feature three Slater-type orbitals for the description of valence electrons. An example is the 6-311G basis set, which uses three Slater-type orbitals for the description of the valence electrons, one expanded in a series of three Gaussians and the other two approximated by one Gaussian each.

The larger the basis set, the lower is the predicted energy of the system, and thus the closer to the real energy. However, the optimized geometries predicted by minimal basis sets are sometimes no worse than those predicted by double-zeta basis sets. For instance, the double-zeta basis sets predict too-large bond angles for water, ammonia, and the HOC angle in alcohols. The minimal basis sets predict that these angles will have values that are too small but closer to the experimental values than the ones predicted by the double-zeta basis sets. Energy differences and reaction energies are predicted better by double-zeta basis sets than by the minimal basis sets.

In order to improve even further the results obtained through the use of Gaussian basis sets, polarization functions are introduced. These are d functions on nonhydrogen atoms and p functions on hydrogens. Polarization functions possess angular momentum beyond that required for the ground state of the atom, while split-valence basis sets allow the orbitals to change size but not shape. The use of polarization functions increases greatly the accuracy of the results, especially where the bond angles are concerned. An even greater improvement due to polarization functions is observed in the prediction of the puckering of rings. This problem will be discussed in more detail in the next chapters. Basis sets containing polarization functions predict values too short for certain bond lengths. This problem is remedied by using them in conjunction with correlation energy calculations, as will be shown later.

For species rich in electrons, such as anions, it is advisable to add diffuse functions to the basis set in order to provide a better description of the system. Such basis sets, for instance 6-31+G*, add diffuse s- and p-type functions to nonhydrogen atoms, while the 6-31++G* set also adds p functions to the hydrogen. Negatively charged amino acids, such as aspartic and glutamic, are particularly prone to requiring the use of diffuse functions.

Larger basis sets make use of more than one d function and of f functions, such as 6-311G* ($2df,2pd$), which uses two d functions, or basis sets with 3 df, which use three d functions besides the f.

Atomic Electrical Charges

Parameters of great importance for the description of a molecule are the electrical charges on each atom. These are of particular interest when the system to be studied is an amino acid or peptide molecules that can be neutral or charged or that exhibit the structure of a zwitterion. Two of the methods to evaluate the net atomic charges are Mulliken population analysis and the Merz–Kollman–Singh method. The Gaussian programs use the former as default and the latter if the command Pop = MK is given.

Mulliken population analysis calculates the total atomic charge on an atom X as the atomic number of X minus the gross atomic population expressed as the sum of the net population of the functions associated only with atom X and half of the overlap population of the functions associated with both atom X and any atom bound to it. This method uses the concept of electron density functions.

The Merz–Kollman–Singh method fits the electrostatic potential to points selected on a set of concentric spheres around each atom.

Two other methods besides the Merz–Kollman–Singh that are used to select the points where point charges are assigned to fit the computed electrostatic potential are CHelp and CHelpG.

Another method to obtain atomic charges, natural population analysis, is carried out in terms of localized electron pairs that act as bonding units.

An example of the difference between the net atomic charges predicted by Mulliken population analysis and by the Merz–Kollman–Singh method can be observed in the charges obtained for one of the conformations of glycine. This conformation does not feature hydrogen bonds and sets the N–C–C–O atoms in the same plane, as shown in Figure 1.1. The optimization was performed at the Hartree–Fock level, using the 6-31G* basis set. The following results were obtained for the net atomic charges (the units are eu):

Atom	Mulliken	Merz–Kollman–Singh
C1	−0.215	0.349
C2	0.748	0.765
N	−0.838	−1.090
O1	−0.702	−0.735
O2	−0.550	−0.583
H1	0.208	0.044
H2	0.178	−0.043
H3	0.468	0.487
H4	0 345	0.409
H5	0.358	0.397

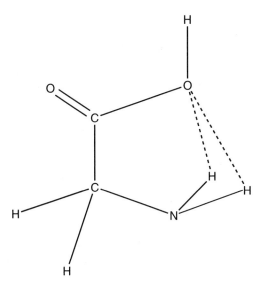

FIGURE 1.1 A conformer of glycine.

The basic difference between the two sets of charges is, as can be seen, the charge on C1 and the hydrogens attached to it. While the Mulliken population analysis method predicts a high polarity to the C–H bond, with the negative charge set on the carbon and a positive charge on the hydrogen, for the same bond the Merz–Kollman–Singh method predicts the hydrogen to be almost neutral and sets the positive charge on the carbon. The negative charge is significantly set on the nitrogen attached to the same carbon.

However, a set of calculations on acetylene, at HF/6-311G** optimized geometry, features the predicted by Mulliken population analysis charges as −0.129 on the carbons and 0.129 on the hydrogens. The Merz–Kollman–Singh method increases the charge separation to −0.302 on the carbons and 0.302 on the hydrogens. If the positive ion of acetylene is investigated by the same method in the Mulliken population analysis case, the charge of 1 is spread almost equally among the carbons and the hydrogens (0.237 and 0.263, respectively), while the Merz–Kollman–Singh method sets more charge on the hydrogens (0.167 on the carbons and 0.333 on the hydrogens). Therefore, it is hard to derive general conclusions about the trend of the differences between the two methods.

Post-Hartree–Fock Methods

Even though the Hartree–Fock method describes well the ground state of atoms and molecules, the wave functions used do not contain a term that would take into consideration the fact that electrons repel each other due to their negative charge. To give the electrons enough space to get away from each other, they should be allowed to make use of energy levels that are not occupied in the ground state. Mixing terms describing excited states into the wave function accomplishes this goal. Instead of using a function consisting of a single determinant, the wave function is described by a linear combination of determinants in which each one represents a particular electronic configuration. This type of wave function is said to represent a configuration interaction. The energy obtained with such a wave function leads to the computation of a lower energy than that obtained with the single-determinant wave function. Since the former includes the electron correlation, the difference between the two energies is called "correlation energy."

In some systems it is mandatory to include the correlation energy. Such systems include chemical reactions in which a bond is broken or formed. Indeed, the correlation energy varies greatly from that of a pair of electrons forming a bond to that of the electrons separated by the breakage of the bond.

An investigation that absolutely requires the calculation of the correlation energy effects is the evaluation of the binding energy of hydrophobic Van der Waals complexes. The main component of the binding in these cases is the dispersion forces, which are taken into account via correlation energy. As an example, one can look at calculations of the binding energy of 4-mercapto pyridine as a fragment of the AG337 (Thyseq) antitumor drug, which is an inhibitor of the enzyme thymidilate synthase. This fragment interacts with the isoleucine residue in the wild-type enzyme and with other amino acid residues in mutant enzymes, where Ile 108 is replaced by other amino acids. The binding energies of the bound complexes are less than one Kcal/mole at the Hartree–Fock level but attain values of around 2 Kcal/mole when the correlation energy is taken into account (1).

The same situation arises for the description of aromatic–aromatic interactions, as related to binding of aromatic substrate fragments to such amino acids side chains as phenylalanine, tryptophan, or tyrosine, in the hydrophobic regions from the interior of proteins. This kind of binding is found, for instance, in the inactivation of chymotrypsin by 6-chloro-2-pyrone compounds, or in the binding of methotraxate (anticancer drug) to dihydrofolate reductase. A study of the latter (2) has found that in order to obtain results similar to those obtained by other methods, the ab initio calculations have to include the correlation energy.

The estimation of the correlation energy of a system can be obtained by a number of methods.

The most accurate method, full configuration interaction, requires a large computational effort and as such is difficult to be applied to relatively large systems. Instead, one uses limited configuration interaction, such as CID, which includes only double excitations. A better correlation energy is obtained if both double and single excitations, such as in CISD, are used. Higher CI are now available, and they can be applied to small amino acids.

Another way to compute the correlation energy is a perturbation method called Moller–Plesset (3). Treating the correlation as a perturbation leads to an expression for the Hamiltonian of the system,

$$H = H^o + \lambda H^1,$$

where H^o is the noncorrelated Hamiltonian operator and H is the total Hamiltonian operator of the system; λ is a parameter defining the perturbation. The exact wave function and energy are expanded in powers of λ. Taking into account various powers of λ, one computes MP2, MP3, and MP4 terms. The higher terms one uses, the better is the value of the correlation energy. However, even for large systems for which only MP2 can be used, the results represent a significant improvement over the Hartree–Fock results.

The Gaussian Programs

There are a number of programs that perform ab initio calculations. Among these, the Gaussian series of programs are the product of Gaussian Inc., Pittsburgh, Pennsylvania.

Gaussian programs are used to perform calculations on many biological systems. They are equipped to handle amino acids as well as small polypeptides. The quantum-chemical calculations applied to such systems can be semi-empirical (such as AM1, MINDO/3, or PM3) or ab initio calculations.

At present, versions of Gaussian 94 and 98 are available for personal computers (Gaussian 94W, where the W stands for Windows). The output produced by this version is identical to that of the workstation and supercomputer versions.

In order to apply the Gaussian program to the study of a system, one has to prepare an input file, to run the program, and to examine and interpret the results.

The input consists of the specification of the molecule or atom to be studied in terms of its electrostatic charge, spin multiplicity, and geometry. In addition, the basis set used and the method (Hartree–Fock or post-Hartree–Fock) have to be specified. By default, the Mulliken population analysis method is used to calculate the net atomic charges. If another method is desired, it has to be stated by the command POP = MK, for instance, if the desired method is the Merz–Kollman–Singh method, or POP = NBO if the natural orbitals method is applied. This command is set on the same line with the specification of the method and of the basis set. For instance, suppose one wants to use the minimal STO-3G basis set, at the Hartree–Fock level, performing geometry optimization and using the Merz–Kollman–Singh method for the calculations of the net atomic charges. The line will read: HF/STO-3G opt POP = MK.

The line specifying the charge and multiplicity reads: 0   1 for a neutral species, which is a singlet. A positively charged monocation that features a spin of $\frac{1}{2}$ will read: 1   2, since the spin multiplicity is $2s + 1 = 2$ for $s = \frac{1}{2}$.

After the method, basis set, charge, and spin have been specified, the initial geometry for the system has to be described, using either x, y, z coordinates for each atom or a Z-matrix that describes the interatomic bond lengths, bond angles, and dihedral angles. For a single-point energy calculation, this geometry is used. For a geometry optimization, each geometrical parameter is modified, and the derivative of the total energy of the system with each parameter is set to zero, in order to obtain the value of the geometrical parameters that will provide a minimum value for the energy. Since a zero value for the first derivative also provides the maximum values for the function examined, in this case the energy, the values of the diagonalized matrix of second derivatives have to be positive. A negative value for any of these indicates that the state of the system is not a minimum but is perhaps a transition state.

An example of a Z-matrix, for the smallest amino acid, glycine, is

```
C
C   1 CC
N   1 CN    2 CN
O   2 CO    1 OCC    3 OCCN
O   2 CO1   1 OCC1   4 180.0
H   4 OH    2 HOC    5 HOCO
H   3 NH    1 HNC    2 HNCC
H   # NH    1 HNC    7 HNH     1
```

The 1 present in the last line, at the end, indicates that the angle HNH is a bond angle and not a dihedral angle.

For the basis set 6-311G*, which is a triple zeta set, using three Slater orbitals for the valence electrons, one expanded in a series of three Gaussians and the other two approximated by a Gaussian function each, while for the core electrons one Slater orbital is used, expanded in a series of 6 Gaussians, plus d polarization functions for the nonhydrogen atoms at Moller–Plesset second-term level with geometry optimization, the route card reads

MP2/6-311G*   opt

The charge and spin multiplicity card reads

   0   1

A card providing the name of the job has to be included, in this case,
Glycine
The order of the cards is
MP2/6-311G* OPT
Blank
Glycine
Blank
0   1
Z-matrix

After the Z-matrix, another blank card is included. The initial values of the
parameters used in the Z-matrix have now to be specified. Unless there is
prior knowledge of these values, empirical values are used. For instance, a
single CC bond is set to 1.54 A, a CN bond length to 1.4 A, and so on. If
some of these values are to be optimized while others are to be kept frozen
throughout the calculation, the former are typed first, followed by a blank
card, which is followed by the parameters that have to be kept at the initial
value. For instance, if one wants to optimize the water bond lengths but not
the HOH angle, the input will read

0   1
O
H   1 OH
H   1 OH   2 HOH
    OH = 0.95
blank
    HOH = 104.5
blank

When the job is terminated and the output can be examined, the en-ergy
of the system at the Hartree–Fock level is read as :scf done : E(RHF)
= .... The RHF stands for "restricted Hartree–Fock" calculations, which
are applied to closed systems. An open system (a system with an unpaired
number of electrons) will be investigated with the UHF method (unre-
stricted Hartre–Fock). If the goal of the calculation included geometry
optimization, a table will contain the optimized geometrical parameters.
These parameters, as mentioned before, correspond to a state of minimum energy. However, it is possible that this minimum is a local minimum
and that there are other minima, including the state featuring the lowest
energy possible, which is the global minimum. In general, local minimum
problems do not arise for bond lengths and bond angles, since these
usually feature only one optimum value. Dihedral angles, though,
which characterize different conformations, could adopt very different
values for different local minima. Therefore, the whole range of possible
conformations has to be examined. For instance, suppose that a dipeptide

is geometry optimized. The angle omega has to be set at 0.0 degrees for one set of calculations and at 180.0 degrees for the other. The program is not going to choose the lowest energy conformation by itself; both conformations have to be investigated separately. For such dihedral angles as psi and phi, again different initial geometries have to be tried until the global minimum is found. Some conformations, even though found to be stationary points, are not real minima. This is illustrated, as mentioned before, by the fact that there are negative values among the eigenvalues of the second-derivative matrix (called the Hessian matrix). Sometimes, though, this criterion is not reliable enough, since these values are not calculated analytically. In consequence, a more reliable criterion is to examine the vibrational frequencies of the system and search for imaginary values.

Frequencies are calculated when the keyword "freq" is added to the route card. They are valid only for optimized structures. The calculation of frequencies also must use the same basis set as the one employed for the geometry optimization.

In addition to the frequencies, the Gaussian programs compute the infrared intensities and the Raman depolarization ratios and scattering activities for the spectral lines.

The frequencies are used to calculate the zero-point vibrational energy, which accounts for the effect of the molecular vibrations at zero kelvin. This energy has to be added to the total energy of the system. When energies of binding of subsystems are calculated for a complex A–B, the difference between the zero-point energy of the complex and the sum of the zero-point energies of A and B has to be added to the binding energy. Of course, if either A or B is an atom, its zero-point vibrational energy is zero.

Besides the zero-point vibrational energy, the keyword "freq" leads to an analysis of the system from a thermochemistry point of view. Unless otherwise required, the temperature is 298.15 K, and the pressure is set at one atmosphere. The main isotopes are used for the atoms of the system.

Here is an example of calculations for one of the conformations of glycine, shown in Figure 1.1.

This conformation features a bifurcated bond between the oxygen of the carboxyl that has the hydrogen and the two hydrogens of the nitrogen. The OH hydrogen is directed away from the nitrogen, kept at 180.0° with it. A geometry optimization is performed, using the 6-31G* basis set. An energy of −282.828066 au (atomic units, where an atomic unit = 627.5 kcal/mole) is obtained. The optimized parameters have the values

C1C2 = 1.5197
C2O1 = 1.3297
C2O2 = 1.1879

OH = 0.9528
C1H = 1.0839
NH = 1.000
NC1 = 1.4404
O1C2C1 = 113.934
O2C2C1 = 123.556
NC1C2 = 118.281
H1O1C2 = 107.857
HNC1 = 111.228
HNH = 106.895
HC1C2 = 106.259
HNC1O1 = 59.406

These parameters are measured in angstroms for the bond lengths and in degrees for the bond angles and for the dihedral angles.

Examining the eigenvalues of the Hessian matrix (the matrix of second-derivatives), one finds only positive values, giving the impression that this structure is a real minimum. The molecule is then subjected to a frequency and thermochemistry calculation, with the "freq" command inserted in the route card.

The frequencies obtained contain one imaginary frequency, which means that this conformation is not a real minimum, but rather a transition state. This frequency is ignored by the program when the zero-point vibrational energy is calculated and found to be 54.335 kcal/mole. As mentioned before, to obtain a more reliable value, this number has to be multiplied by a value of about 0.89, at Hartree–Fock level.

A heat capacity of 15.350 kcal/mole K is also calculated, as well as an entropy of 69.431 kcal/mole K. The thermal energy correction is 0.091 au, leading to a corrected total energy of −282.736969 au. The temperature is 298.15 K.

In addition, the program separates the above-mentioned values into translational, rotational, and vibrational contributions.

Besides the Gaussian programs, there are other programs, such as Gamess (see Chapter 4), used to perform quantum-chemical calculations.

The following chapters will examine quantum-chemical calculations as applied to different amino acids and small peptides.

References

1. Sapse, A.M., Sapse, D., Tong, Y., and Bertino, J.R. *Cancer Investigation.* To be published.
2. Sapse, A.M., Schweitzer, B.S., Dicker, A.P., Bertino, J.R., and Frecer, V. *Int. J. Pept. Prot. Res. 39*, 18, 1992.
3. Moller, C., and Plesset, M.S. *Phys. Rev. 46*, 618, 1934.

Bibliography

Foresman, J.B., and Frisch, A. *Exploring Chemistry with Electronic Structure Methods.* Gaussian Inc., Pittsburgh, PA, 1993.

Hehre, W.J., Radom, L., Schleyer, P.V.R., and Pople, J.A. *Ab Initio Molecular Theory.* John Wiley & Sons, New York, 1986.

Levine, I.N. *Quantum Chemistry.* Prentice Hall, Englewood Cliffs, New Jersey, 1991.

Schaefer III, H.F. *Applications of Electronic Structure Theory.* Plenum Press, New York and London, 1977.

2
Theoretical Calculations on Small Amino Acids

The smallest amino acid, glycine, was among the first amino acids to be investigated by theoretical methods. Its small size makes it particularly suitable for ab initio treatment, even with large basis sets. The results of the ab initio geometry optimization of glycine (1) first suggested the existence of a state undiscovered by microwave spectroscopic studies. This state is the S (extended) state of glycine. Since only the C state (Figure 2.1) had been observed experimentally, it was concluded that the theoretical results were at fault (2). Accordingly, Sellers et al. (3) and Schafer et al. (4) used the gradient method of Pulay (5), with the 4-21G basis set at Hartree–Fock level and found indeed the S structure to be more stable than the C. This theoretical result was confirmed experimentally by the microwave study of Suenram and Lovas (6).

A number of studies have applied ab initio calculations to the investigation of glycine (7–9), while other studies have applied semi-empirical methods to the same system (10–11). Ramek, for instance, used the Gaussian 80 and the Gamess programs with the 4-31G basis set to study a number of amino acids, including glycine and alanine (7). That study suggests that the strength of the intramolecular O–H . . . NH_2 bond increases with the size of the ring, due to the larger number of degrees of freedom available to larger rings. Therefore, the fact that the extended conformation of glycine is more stable than the form that exhibits a hydrogen bond between the hydrogon set on the oxygen and the lone pair of electrons on the amino nitrgen is explainable by the size of the ring formed. However, according to Ramek, the C structure is the kinetically most stable conformation, with a rotational energy greater than or equal to 43 kJ/mole for all internal rotations in the potential energy surface of glycine. Since it takes a large energy to destroy C once formed, it has a longer lifetime than other conformations and can thus be easily detected by spectroscopic methods.

These studies (7–9) explain that the microwave spectrum favors the structure C because of its higher dipole moment. Indeed, the intensities of transitions in the microwave spectrum are proportional to the square of the dipole moment components.

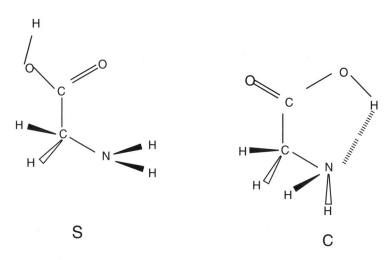

FIGURE 2.1 Extended (S) and cyclic (C) conformers of glycine.

As shown by Ramek et al. (7), the 4-21G calculations of both C and S structures reproduce accurately the microwave rotational constants. Microwave research for S led to the discovery of some weak transitions close to the predicted values, and thus S was also found experimentally to be more stable than C.

The exact shape of the energy surface around C has been dicussed in (16–17) and it has become apparent that the level of the calculations strongly influences the results.

The Gaussian 86, the Gamess, and the Texas (12) programs have been used to study several conformations of glycine, including S and C. Of the three conformations other than S and C, two set the oxygen atom to be not coplanar with the two carbons and the nitrogen, one of them with the OH hydrogen directed toward the nitrogen, the other away from it. The third one has the four atoms coplanar and the OH hydrogen away from the nitrogen.

In addition to the 4-21G basis set, the study (7) uses the UQ10 basis set (13) and Dunning's basis set (14), as well as larger double-zeta basis sets with or without polarization functions and diffuse functions.

When the planar and nonplanar hydrogen-bonded conformations are compared, it is found that as the size of the basis set increases, their energies become higher than that of the global minimum S (from 7.65 kJ/mole for the planar conformation and 5.81 kJ/mole for the nonplanar one for the STO-3G basis set to 11.73 kJ/mole and 11.28 kJ/mole, respectively, for the 6-31++G** basis set). At the same time, the difference in energy between these two structures decreases with the size of the basis set, going from −1.84 kJ/mole at the STO-3G level, to −0.44 kJ/mole at the

6-31++G** level, with the nonplanar conformation being more stable than structure C when minimal basis sets and basis sets with polarization functions are used. Ramek et al. (7) suggest that this stability is due to the fact that the twisting of the NH_2 group reduces the strain introduced by the bond eclipsing. Significant twisting, though, leads to nonplanar moments, in disagreement with observed values (15). When the structure C was optimized without constraints, it was found that it represents a saddle point, not a minimum, and the optimized NCCO angle differs from 180.0°. Simple double-zeta basis sets, as well as the 10s, 5p/4s sets, indicate structure C as more stable. It can be seen, thus, that for a proper description of glycine, polarization functions are necessary when double-zeta basis sets are to be used. The structures featuring the OH hydrogen away from the nitrogen and the NH_2 hydrogens directed toward the oxygen atom also depend for their description on the basis sets used. With a fairly large basis set (6-31+G**), the difference between the planar and the nonplanar structures becomes negligible, only about 0.01 kJ/mole. The authors conclude that for a proper description of the potential energy hypersurface of glycine it is necessary to use large basis sets, with correlation energy calculations included. As such, they report MP2/6-311G** calculations, which establish structure S as the most stable, followed by the nonplanar structure with the OH hydrogen directed to the nitrogen atom, higher in energy than structure S by 2.92 kJ/mole and with the structure C higher than S by 3.53 kJ/mole.

A later study on glycine and N-formylalanineamide (16–17) examines the importance of correlation-gradient geometry optimization for conformational analyses. Although the authors recognize the fact that such methods as CI and MCSCF are difficult to apply to larger systems, they suggest that perturbative methods such as Moller–Plesset, even if only at the MP2 level, in combination with large basis sets, include a large part of the correlation energy. Since the NH...O and OH...N interactions are particularly sensitive to dispersion effects, it may become important to examine Hartree–Fock optimized geometries in comparison with MP2 optimized geometries. Frey et al. (16) determined the geometries of three forms of glycine: Two of them feature the N–C–C–O atoms in the same plane, one with a NH...O and the other with N...HO hydrogen bonds, and the third has the oxygen out of the NCC plane and also features an OH...N hydrogen bond. It was found that the nonplanar form is favored by MP2 calculations, in disagreement with the microwave experiment in which the second planar form is observed. The HF calculations contain the 4-21G basis set and the triple-zeta 6-311G**, which contains d functions on the nonhydrogen atoms and p functions on hydrogens. The MP2/6-311G** was used to examine the effect of the correlation energy on the geometry. The HF results are similar to the one discussed above. The MP2/6-311G** geometries lead to differences in energy shifted by 50%–60% from the differences predicted by MP2 calculations with the HF geometry (MP2//Hf/6-311G**). It seems thus that MP2 geometry optimizations might be necessary in the glycine case.

A set of calculations on glycine, using the HF/6-311+G** basis set, which is a triple-zeta set with diffuse functions and polarization functions of the *d* type on nonhydrogen atoms and of *p* type on hydrogen atoms, was performed for the conformations of glycine shown in Figure 2.2. In addition

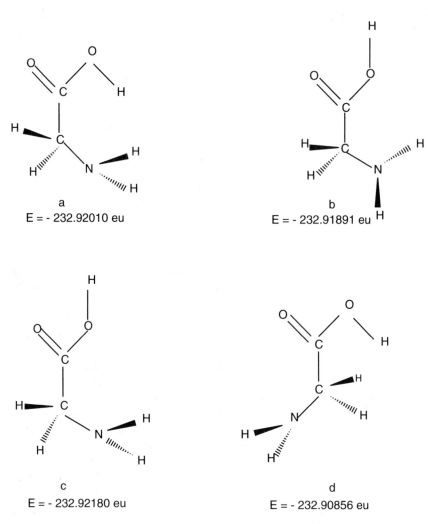

FIGURE 2.2 a. Glycine conformer with hydrogen bond between the carboxyl's hydrogen and the nitrogen. b. Glycine conformer with the N slightly above the CCO plane. c. Glycine conformer with a hydrogen bond between NH and OH. d. Glycine conformer with N antiplanar to the CCO plane.

to Hartree–Fock calculations, the correlation energy was estimated by the Moller–Plesset perturbation method, up to the MP4 term. The Moller–Plesset calculations were performed in order to obtain single-point energies at the Hartree–Fock optimized geometry.

As can be seen in Figure 2.2, the most stable of these conformers is 2.2c, which features a hydrogen bond between the nitrogen's hydrogens and the oxygen of the carboxyl. This hydrogen bond features a rather long distance between the oxygen and the hydrogens, 2.7Å, with the two hydrogens set above and below the NCCO plane. The next in order of stability is the structure 2.2a, which features a hydrogen bond between the hydrogen set on the carboxyl's oxygen and the nitrogen atom. The distance between this hydrogen and the nitrogen atom is 2.04 Å, a value more typical for a hydrogen bond. The energy difference between these two structures is only 4.5 kJ/mole. The third in stability, structure 2.2b, features the nitrogen atom slightly above the CCO plane, forming a dihedral angle of 15°. The distance between the oxygen and one of the nitrogen's hydrogens is 2.4 Å. The OH hydrogen is positioned at 180.0° to the nitrogen. This structure is higher by 7.6 kJ/mole. Finally, structure 2.2d, which features the nitrogen antiplanar to the CCO plane, is the least stable out of the four, with an energy higher by 34.7 kJ/mole. The energies discussed above are Hartree–Fock energies.

The correlation energy effect on the stabilities of these conformers as estimated by the Moller–Plesset method is as follows:

2.2a   MP2 = −283.78313, MP3 = −283.79480, MP4 = −283.80961,
2.2b   MP2 = −283.77927, MP3 = −283.79136, MP4 = −283.80637,
2.2c   MP2 = −283.78236, MP3 = −283.79429, MP4 = −283.80839,
2.2d   MP2 = −283.77127, MP3 = −283.78335, MP4 = −283.79826.

It can thus be seen that when the correlation energy is taken into account, the order of stability changes slightly, with 2.2a the most stable conformer, while 2.2d remains the least stable conformer, followed by 2.2b, similar to the Hartree–Fock results. The relative energies, taking 2.2a to be zero, are as follows (in kJ/mole):

	MP2	MP3	MP4
2.2a	0	0	0
2.2b	10.12	9.0	8.5
2.2c	2.0	1.3	3.2
2.2d	31.1	30.0	29.7

As can thus be seen, besides changing the order of stability, the differences between the energies of various conformers become smaller as higher-order terms are used, with the exception of conformers a and c, for which MP4 shows the higher relative energy.

The role of polarization functions in the determination of the structure of glycine is also dicussed by Ramek and Cheng (17). This study states that while the 4-31G calculations result in the finding of seven symmetry-unique local minima for neutral glycine, the 6-31G* calculations show the presence of eight symmetry-unique local minima. There seems to be a perfect match between the 4-31G and the 6-31G* obtained conformers for two of them that feature low energy. However, the second-lowest energy, which features a 1.9 kcal/mole relative energy, is planar for the 6-31G* set but twisted for the 4-31G set, with the opposite true for the conformer featuring a relative energy of 2.9 kcal/mole.

The additional minimum found with the 6-31G* calculations features the N cis to the OH group and a clustering of the 3 hydrogen atoms, one of the OH group and the other two of the NH_2 group. If the NH_2 group is inverted, this structure reverts to a conformer found also to be a local minimum by the 4-31G basis set calculations and that features a hydrogen bond between the OH hydrogen and the nitrogen atom. This structure is similar to the one pictured in Figure 2.2a. As shown by Ramek and Cheng (17), the inversion energy takes a value of 0.33 kcal/mole with 6-31G* calculations. They also show that a number of calculations show that for the hydrogen-clustered conformer to be a minimum, polarization functions are necessary.

Studies of Beta-Alanine

Alanine is the simplest beta amino acid, and it enters the metabolic mainstream at the pyruvate entry point. Its transamination yields pyruvate via the reaction

Alanine + alpha-Ketoglutarate = pyruvate + glutamate.

Beta-alanine is present in insect cuticle; animal muscles, liver, and brain; and plants (18–20). It also acts as an inhibitor of the nervous system (21–23).

Ramek reported an extensive study of neutral beta-alanine (24). In a subsequent paper (25) some results of that study are discussed.

Ab initio self-consistent-field calculations were performed on beta-alanine, using the Gaussian-80 and the Gamess computer programs. The 4-31G basis set was used.

Ramek (25) found 38 local minima on the potential-energy surface of the neutral amino acid, with two of those being mirror-symmetric conformations. The other 36 conformers are divided into 18 pairs of mutually mirror-symmetric entities. Those with positive dihedral C–C–C–N angles are labeled I, II, etc., while those exhibiting a negative C–C–C–N angle are labeled Im, IIm, etc. The numbering system labels C1 the carboxylic carbon and C3 the amino-bearing carbon, with C2 in the middle. Among the different conformers representing local minima, two groups appear: In one, the relative energies are below 12 kcal/mole, being thus relatively low, while the second group features relative energies from 31 to 48 kcal/mole, constituting a less stable group of conformers. The global minimum of all the conformers features a near coplanarity of the COOH group with the CCC plane. The nitrogen atom is out of the plane, with a dihedral angle of about 60°. This makes the two CH2 groups adopt a staggered conformation. When the relationship between the energy of the different conformers and their structures is examined, Ramek points out several trends: With respect to the planarity of the COOH group, with the OH and CO either in cis or trans conformation, the most stable conformations feature the COOH group coplanar with the CCC plane, while the less stable conformations feature a dihedral angle between these two groups of about 120°; the NCCC backbone is in a transplanar conformation or in a gauche orientation. The low-energy group is characterized by a zero value for the HOCO angle, while the high-energy group affords for it a value of 180°, with the exception of a conformer with low energy and a value of −177.6° for this angle. The stability of this conformer is due to the formation of a hydrogen bond between the nitrogen and the OH hydrogen. Indeed, the distance between the two atoms is only 1.864 Å, a typical distance for a hydrogen bond.

Ramek reports (25) a study of the reaction paths between different conformers, consisting of rotations of the COOH group, of the OH group, of the NH_2 group, and of the inversion of the NH_2 group.

The intramolecular OH . . . N hydrogen bond plays an important role in the energetics of the interconformer conversion reactions. A coupling of the COOH group rotation and of the OH rotation that preserves the hydrogen bond is induced. In the reactions that lead to the irreversible formation of the conformer featuring the OH . . . N hydrogen bond, denoted by V, a number of rotations might lead to the reaction path V = Vm.

Another strong intramolecular hydrogen bond is formed between the hydrogen atom on the OH and the carbonyl oxygen. This interaction is present only in conformations with the cis orientation of the COOH group. This interaction lowers the energy of the conformer by about 40 kJ/mole and increases the O–H bond length by approximately 0.005 A. The C=O bond length is also increased, by approximately 0.007 A. The C–O and CC–COOH bond lengths are decreased by 0.005 and 0.007 A, respectively. The C=O and the O–H stretching vibration frequencies also decrease.

The O–H . . . N hydrogen bond features a stabilizing effect of about 38 kJ/mole. The other possible hydrogen bonds, N–H . . . O=C and N–H . . .

O–C, are weak interactions, which do not influence bond lengths or frequencies.

However, the global minimum contains this interaction, as well as the O–H . . . O=C hydrogen bond, thus showing that the sum of the two is more energetically favorable than the presence of the O–H . . . N hydrogen bond alone.

Heal et al. (26) analyzed a family of 20 stable conformations of beta-alanine using the topological shape group method (27). The topological shape group method is a general molecular shape analysis technique using the three-dimensional electronic density, calculated in this case through the use of the 6-31G** basis set. The basis of this method is the existence of shape-property relations. The method determines the shape information for the chemically relevant range of electron density and analyzes shape in detail. The limitations of the method are set by the quality of the electron density and by the resolution set by the user (27).

Heal et al. (26) use the method on the 20 conformational isomers of beta-alanine with unique intrinsic geometry, as obtained via ab initio calculations by Ramek (25). The energies and electronic density distributions were recalculated with the 6-31G** basis set, as implemented by the Gaussian-92 (28) computer program.

The change of basis set from the 4-31G used by Ramek (25) to the 6-31G** basis set used by Heal et al. (26) causes changes in the order of some of the conformers' stability. The conformer labeled V, which features an O–H . . . N hydrogen bond, is stabilized by the use of polarization functions.

The two conformers lowest in energy that contain a N–H . . . O=C hydrogen bond are more separated in energy by the use of polarization functions.

Heal et al. (26) show that the quality of the basis set used for obtaining the electron density affects the analysis. However, for double-zeta and larger sets there is little change in the shape of the electron density within the chemically relevant range. Thus, optimizing the geometries at 4-31G level and using a larger set, such as the 6-31G** basis set, for the calculations of the energies and recalculations of the electron densities is sufficiently accurate. For instance, the 6-31G** shape group analysis provided excellent correlations between the shape and the toxicological activity in a family of polycyclic aromatic hydrocarbons (29).

The changes in intramolecular interactions are monitored well by similarity indices, which are a good measure of similarity between various conformers. They are also a reliable tool for the identification of outstanding isomers, and they are very sensitive to changes in nonbonded interactions.

A microwave spectroscopic investigation (30) in which two conformations of alanine were found in gas phase inspired Cao et al. (31) to perform high-level ab initio studies on alanine. They consider alanine a particularly important molecule because it is the smallest amino acid with a chiral alpha-carbon atom, and many conformational studies of peptides and proteins can use it as a model.

Cao et al. (31) use the 6-31G** and the MP2/6-31G** basis sets as implemented by the Gaussian-92 computer program to geometry optimize the structures of 13 conformers of alanine. In addition, they perform geometry optimization for the three more stable conformers using the 6-311G** basis set, at Hartree–Fock level and MP2 calculational level. A number of single-point calculations were performed, using the MP2/6-31G** geometry, using such basis sets as 6-311+G (2d,2p), 6-311+G (3df,2p) at MP2 level, and smaller basis-set (6-31G**) calculations, and using the MP3 and MP4 Moller–Plesset terms as well as QCISD calculations.

When the Hartree–Fock geometries are compared to the geometries obtained via the inclusion of the correlation energy effects, it is possible to observe important features due to the inclusion of the electron correlation. One of these is bond-length elongation, of 0.03 A for C–O, C=O, and H–O, and 0.01 A for N–H and C–N. In some cases, an elongation of 0.005 A is found for C–C and C–H. The bond angles are contracted by 2°–3° for H–O–C and H–N–C angles, whereas the H–C–C and the O–C–O angles expand with correlation. The distance between the carboxyl's hydrogen and the other oxygen is remarkably constant at all computational levels, about 2.26 A for the cis conformers and 2.97 A for the trans. However, in the conformers that feature a hydrogen bond between this hydrogen and the nitrogen atom, the distance changes from 2.03 A at Hartree–Fock level to 1.9 A at MP2 level. In general, as expected, absolute values are more sensitive to the computational method than relative trends (31).

The rotational constants and dipole moments as calculated by Cao et al. (31) are in good agreement with the two similar conformers identified in gas phase by Godfrey et al. (30), especially at high computational levels. Comparing their results with those obtained for N-acetyl-N′-methylalanine amide (32), Cao et al. (31) conclude that the conformational properties of monomer amino acids conform to their structural functions as residues in peptides and proteins.

Besides the energetics and geometry of glycine and alanine, researchers have looked into the affinity of glycine and glycine methyl analogues for cations, such as H+, Li+, and Na+. Jensen (33) and Bouchonnet and Hoppillard (34) apply molecular orbital calculations to evaluate the binding energy between glycine and each cation. Indeed, such information is of great value since one of the new approaches for obtaining structural information about peptides is the complexation with metals (35).

Bouchonnet and Hoppillard (34) use the MP2/6-31G*//HF/3-21G method of calculation to investigate the complexes of glycine with H+ and Na+. They found a total of three structures corresponding to protonated glycine and five structures for the glycine–sodium complexes.

Jensen (33) uses the 6-31G* basis set with the GAMESS program package. The protonated glycine minima and some of the metal structures were characterized by calculating the vibrational frequencies. The correlation energy effects were taken into consideration by performing MP2/6-31G*//HF/6-31G* calculations. MP4 calculations using the 6-31G* basis set and MP2 calculations using the 6-31+G (2d) basis set were performed for the five lowest energies of glycine-Li+. The Moller–Plesset calculations were performed with the Gaussian-90 program, and MNDO calculations were performed with the MOPAC program (36).

In addition to the glycine calculations, Jensen (33) performed calculations of the lithium affinities to alanine (by replacing a hydrogen of glycine by a methyl group), to sarcosine (by placing a methyl on the nitrogen), and to the methyl ester of glycine.

Jensen (33) found nine minima with Cs symmetry for the gas-phase protonated glycine. The lowest-energy structure is the one featuring the protonation of the nitrogen of the lowest-energy form of neutral glycine. The addition of Li+ and Na+ to glycine results in 13 minima. These complexes are of five types: the metal bound to both nitrogen and oxygen, to both oxygens, to only the carbonyl oxygen, to only nitrogen as added to the lone pair, and by metalation of the amino group. A total of five Li+ complexes and six Na+ complexes are within 10 kcal/mole of the global minima. Three of those have a five-membered geometry, while the others have the metal bound to one or both oxygens.

The order of stability of the protonated species is similar to the lithium and sodium complexes. The absolute energies of protonation agree well with the experimental results (33).

The protonation energies are 222.7 kcal/mole for glycine, 226.4 kcal/mole for alanine, 226.1 kcal/mole for the ester, and 230.3 kcal/mole for sarcosine. These values are obtained via MP2/6-31G* calculations. The values are reduced by approximately 3% when MP2/6-31+G (2d) calculations are performed. The Li+ affinities are 66.4 kcal/mole, 67.7 kcal/mole, 69.6 kcal/mole, and 69.4 kcal/mole, respectively. Again, the introduction of diffuse functions and of two d orbitals reduces them somewhat. In the case of sodium complexes, the MP2/6-31G* calculated values are 46.0 kcal/mole, 47.5 kcal/mole, 48.6 kcal/mole, and 50.8 kcal/mole, respectively. When the 6-31+G (2d) basis set is used, both for lithium and sodium, the values are reduced by about 12% for lithium and 16% for sodium.

The most significant consequence of this work is the fact that it finds structures with different complexation modes to be close in energy, even though the system investigated is a small amino acid. It follows that for peptides a large number of possible modes of complexation have to be considered (33).

References

1. Vishveshwara, S., and Pople, J.A. *J. Am. Chem. Soc.* 99, 2422, 1977.
2. Schafer, L., Newton, S.Q., and Jiang, X. In *Molecular Orbital Calculations For Biological Systems.* Oxford University Press, New York, 1998.
3. Sellers, H.L., and Schafer, L. *J. Am. Chem. Soc.* 100, 7728, 1978.
4. Schafer, L., Sellers, H.L., Lovas, F.J., and Suenram, R.D. *J. Am. Chem. Soc. 102*, 6566, 1980.
5. Pulay, P., Fogarasi, G., Pang, F., and Boggs, J.E. *J. Am. Chem. Soc. 101*, 2550, 1979b.
6. Suenram, R.D., and Lovas, F.J. *J. Am. Chem. Soc. 102*, 7180, 1980.
7. Ramek, M., Cheng, V.K.W., Frey, F.R., Newton, S.Q., and Schafer, L. *J. Mol. Structure (THEOCHEM) 235*, 1, 1991.
8. Palla, P., Petrongolo, C., and Tomasi, J. *J. Phys. Chem. 84*, 435, 1980.
9. Dykstra, C.E., Chiles, R.A., and Garrett, M.D. *J. Comp. Chem. 2*, 266, 1981.
10. Laurence, P.R., and Thomson, C. *Theor. Chem. Acta 58*, 121, 1981.
11. Masamura, M. *J. Mol. Structure 152*, 293, 1987.
12. Frisch, M. et al. Gaussian 86. Carnegie Mellon University, Pittsburgh, PA, 1986.
13. Mezei, P.G., and Csizmadia, I.G. *Can. J. Chem. 55*, 1181, 1977.
14. Dunning, T.H., and Hay, P.J. In Schafer, H.F. III (ed.) *Methods of Electronic Structure Theory.* Plenum, New York, 1977.
15. Suenram, R.D., and Lovas, F.J. *J. Mol. Spectroscopy 72*, 237, 1978.
16. Frey, R.F., Coffin, J., Newton, S.Q., Ramek, M., Cheng, V.K.W., Momany, F.A., and Schafer, L. *J. Am. Chem. Soc. 114*, 5369, 1992.
17. Ramek, M., and Cheng, V.K.W. *Int. J. of Quantum Chem., Quantum Biology Symp. 19*, 15, 1992.
18. Miettinen, J.K. *Ann. Akad. Scient. Fennicae A60*, 520, 1955.
19. Bruun, A., Ehinger, B., and Forsberg, A. *Exp. Brain Res. 19*, 239, 1974.
20. Drabkina, T.M., Shabunova, I.A., Matyushkin, D.P., Gankina, E.S., and Efimova, I.I. *Bull. Eksperim. Biol. Med. 101*, 30, 1986.
21. Hosli, L., Tebecis, A.K., and Filias, N. *Brain Res. 16*, 293, 1969.
22. Sandberg, M., and Jacobson, I. *J. Neurochem. 37*, 1353, 1981.
23. Coquet, D., and Korn, H. *Neurosci. Lett. 84*, 329, 1988.
24. Ramek, M. *Habilitationsschrift.* Technische Universität Graz, Austria, 1989.
25. Ramek, M. *J. Mol. Structure (THEOCHEM) 208*, 301, 1990.
26. Heal, G.A., Walker, P.D., Ramek, M., and Mezey, P.G. *Can. J. Chem. 74*, 1660, 1996.
27. Mezey, P.G. *Shape in Chemistry: An Introduction to Molecular Shape and Topology.* VCH Publishers, New York, 1993.
28. Frisch, M.J., et al. *Gaussian 92.* Gaussian Inc., Pittsburgh, PA, 1992.
29. Mezey, P.G., Zimpel, Z., Warburton, P., Walker, P.D., Irvine, D.G., Dixon, D.G., and Greenberg, B. *J. Chem. Inf. Comput. Sci. 36*, 602, 1996.
30. Godfrey, P.D., Firth, S., Hatherley, L.D., Brown, R.D., and Pierlot, A.P. *J. Am. Chem. Soc. 115*, 9687, 1993.
31. Cao, M., Newton, S.Q., Pranata, J., and Schafer, L. *J. Mol. Structure (THEOCHEM) 332*, 251, 1995.
32. Scarsdale, J.N., Van Alsenoy, C., Klimkowski, V.J., Schafer, L., and Momany, F.A. *J. Am. Chem. Soc. 105*, 3438, 1983.

33. Jensen, F. *J. Am. Chem. Soc. 114*, 9533, 1992.
34. Bouchonnet, S., and Hoppillard, Y. *Org. Mass. Spectrom. 27*, 71, 1992.
35. Grese, R.P., Cerny, R.L., and Gross, M.L. *J. Am. Chem. Soc. 111*, 2835, 1989.
36. Stewart, J.J.P. MOPAC, *QCPE Bull. 10*, 86, 1990.

3
Gamma-Aminobutyric Acid (GABA)

In the last decades there has been a rapid increase in interest in the study of the central amino acid neurotransmitters, especially in the role of γ-aminobutyric acid (GABA) in certain neurological and psychiatric problems.

The pharmacology of some neurotransmitters represents the object of many experimental and theoretical studies. Among those, GABA has been investigated from a structural point of view as well as for its activity in the functioning of the nervous system.

The presence of GABA in the mammalian central nervous system (CNS) was demonstrated over 40 years ago. It fulfills the main criteria for the identification of an inhibitory neurotransmitter: it is synthesized within a limited number of nerve terminals and stored in them, its release from CNS tissues can be induced in vitro by electric stimulations and in vivo by impulses in certain neuronal pathways, the pre- and postsynaptic GABA receptor–ionophore complexes have been identified, and GABA shows a depressant action when applied to single neurons, which mimics the effect of the inhibitory transmitter released after neuronal stimulation. In addition, carrier-mediated membrane transport systems and enzymatic processes for the termination of the neurotransmission process and for the inactivation of GABA have been characterized (1). GABA is present in the cerebellar golgi, basket and stellate cells, hippocampal basket cells, neo-striatal spinal cord, and nigrostriatal and pallidonigral neurons (2–4).

Four types of GABA-mediated inhibitory processes seem to operate in the CNS: postsynaptic, presynaptic, recurrent, and collateral inhibition.

GABA can also feature disinhibition, a fact that may be of importance in brain function. The disinhibition involves postsynaptic contact between two GABA neurons.

GABA plays an important role in neurological and psychiatric disorders. For instance, low levels of GABA have been found in postmortem brain tissues from choreic patients. Parkinson's disease features an imbalance between the GABA and dopamine systems. Moreover, in the substantia nigra from patients with Parkinson's disease, the GABA receptor density is below normal. Reduced GABA uptake capacity was found in sites near seizure foci in epileptics, showing degeneration of the GABA neurons. Schizophrenic patients show decreased GABA activity in certain regions of the brain. When free GABA levels were measured in the cerebrospinal fluid of 74 neurological patients suffering from cerebral cysticorsis, Parkinson's disease, multiple sclerosis, meningeal tuberculosis, viral encephalitis, cerebrovascular disease, and several kinds of distonia, fourfold elevation of GABA was found in cystocerosis and twofold elevation of GABA was found in cerebrovascular disease and viral encephalitis. The other diseases did not show increased GABA levels. It was concluded that the GABA level increases in diseases featuring inflammation and tissular necrosis (5).

In the case of neuronal degeneration, receptors in GABA synapses are pharmacological sites of attack. Structure–activity studies of the conformations of different GABA analogues are a great help for pharmacological studies (6–7).

The GABA receptors are divided into two groups: $GABA_a$ and $GABA_b$ receptors. The former are associated with the function of GABA as the transmitter of pre- and postsynaptic inhibition in the mammalian CNS. Less is known about the latter category of receptors (8), which have been detected in the peripheral nervous system and in the CNS by pharmacological and ligand binding techniques (9–10). Different GABA agonists bind to these two categories of receptor: Baclofen binds to GABA receptors, while isoguvacine, for instance, is a specific GABA agonist. The activation of GABA receptors by baclofen seems to be related to a reduction in the release of some neurotransmitters. Such a mechanism would explain the effects of baclofen in certain types of spasticity (11).

Krogsgaad-Larsen et al. (6–7) developed a series of model compounds for the structural specificities of the $GABA_a$ and $GABA_b$ receptors. It has been sugested (12) that the GABA receptor complex in the spinal cord, which comprises recognition sites for GABA and high-affinity binding sites for such drugs as benzodiazepines, can be very similar or even identical to the receptor complex located in the brain (13). It was found (14) that the binding of H3 labeled benzodiazepine receptor sites called BZD1 and BZD2 is always stimulated by $GABA_a$ recognition-site ligands, but it is differentially stimulated by the depressant (+)-etomidate (15). It was thus suggested that there might exist two $GABA_a$/benzodiazepine receptor complex subclasses: $GABA_a$/BZD1 and $GABA_a$/BZD2. This fact might explain the distinct pharmacological effects of BZD featuring similar structures.

When the level of GABA in the brain diminishes below a certain threshold, convulsions can occur. They cease when GABA is directly administered into the brain (16–17). However, GABA does not cross the blood–brain barrier, and consequently, it is not an effective anticonvulsant agent. It has been suggested (18) that in order to increase the concentration of GABA in the brain, the enzyme GABA aminotransferase, responsible for the catabolism of GABA, has to be inhibited. Silverman et al. (18) synthesized a series of substituted aminopentanoic acids and showed that the halogen-substituted species are potent inactivators of GABA T. The fluorine-containing analogue was found to cross the blood–brain barrier. Another series of GABA analogues were found to act as substrates but not as inhibitors (18). In connection with these studies, it was found that GABA is probably bound to GABA T in an extended conformation (19). Some of the compound designed and synthesized by Silverman et al. (18) are 4-amino-2-(substituted methyl)-2-butenoic acids. When the substituent on the methyl is OH, Cl, or F, the compounds are not time-dependent inactivators of GABA-T but its competitive inhibitors. F, however, produces elimination without inactivation and also undergoes transamination. In order to increase binding, Silverman et al. (18) suggest the incorporation of hydrogen-bonding substituents.

In order to translate the animal models as far as GABA activity is concerned to humans, it is necessary to use a nontoxic, well-absorbed GABA agonist (20). Some compounds considered, such as muscimol, isoguvacine baclofen, or sodium valproate were rejected for such reasons as toxicity, lack of ability to enter the brain, activity only at GABA$_b$ receptors, or inhibition of the GABA catabolism. Bergman (20) developed a synthetic compound, progabide. It is a Schiff base of gamma-aminobutyramide and a substituted benzophenone. This compound had been found to be well absorbed and relatively nontoxic. It was found to be useful in exploring the role of GABA in different human diseases. An advantage of this compound is that it acts like GABA at the GABA receptors and not via GABA-T inhibition or blockage or the reuptake systems. Studies of the action of progabide in the rat brain show it to increase the rate of turnover of norepinephrine, to reduce the rate of turnover of serotonin and to reduce the rate of utilization of dopamine (21).

Although acetylcholine and monoamines are specific neurotransmitters and end products of neural metabolic reactions, GABA is also an intermediate and substrate of other biochemical pathways. Accordingly, neurons that use GABA as substrate are divided into compartments (20).

GABA is synthesized in a reaction catalyzed by glutamic acid decarboxylase. The reactant is glutamate.

GABA is found in the largest concentrations in the brain in nucleus accumbens, temporal and frontal cortices, caudate nucleus, hypothalamus, and substantia nigra (22). There appear to be GABAergic neurons in the cortex (23–24), whose function is that of inhibitory interneuron.

There is evidence that the aspiny neurons that contain GABA exist in the striatum. The three major pathways from the striatum to the pallidum and to the substantia nigra are thought to use GABA as a major neurotransmitter. In addition, there is evidence that GABA provides afferent and interneuronal transmission.

The cerebellum, especially the Purkinje cells, which are its output, contain a large quantity of GABA. However, this issue is still not completely undestood. It seems likely that the stellate and basket cells that envelop the Purkinje cells are GABAergic, a fact supported by the finding of GABA receptors on these cells' bodies and dendrites.

GABA is also found in the spinal cord, with the highest concentration in the dorsal horn in Rexed layers II and the border zone II–III (25).

There is no doubt that the GABAergic system is involved in the pathology of convulsive disorders. Bursts of action potential, revealed in experimental models (20), characterized as paroxysmal depolarizing shift (26), may have to do with increased calcium conductance in membrane channels and its effects on the potassium flux. Is GABA involved in producing seizures? The answer is yes and is based on several facts. One of them refers to the human pyridoxine deficiency syndrome. Pyridoxine is a necessary cofactor for the formation of GABA, and its deficiency leads to the production of seizures (28). Also, direct evidence obtained by neurochemical analysis in patients operated on for seizure control shows lowered local GABA (29). In addition, it was found that agents that are used to treat epilepsy interfere with the GABA functions (29). Among these agents, progabide, discussed above, is an antiepileptic agent that proves to be a GABA agonist (20). It is an effective anticonvulsant, especially its acid metabolite.

Another pathological condition, spasticity, is defined as a syndrome of the upper motor neurons, consisting of increased rigidity, resistance to stretching, and exaggerated deep-tendon reflexes. A number of studies (30–31) indicate that GABA activity is involved in this condition. Progabide acts on the muscle spindle or on the neuron control of the fusiform system, thus confirming the role of GABA in the spinal cord.

GABA in the basal ganglia seems to be involved in movement disorders, even though the mechanism and the extent are not completely understood (20). In Parkinson's disease, loss of GABA receptors appears in the substantia nigra, similar to the loss of dopaminergic cells. After treatment with levodopa, the levels of glutamic acid decarboxylase, previously low, returned to the normal range in substantia nigra, caudate, putamen, and pallidum (32). Also, studies have found that untreated patients with Parkinson's disease feature significantly lower GABA levels in their lumbar cerebrospinal fluid than those of patients treated with levodopa (33–34). Another drug used to control the GABA mechanism in patients with epilepsy is vigabatrin (gamma-vinyl GABA) (35).

The effects of vigabatrin cannot be explained only by the elevation of GABA levels. Indeed (36), vigabatrin also affects beta-alanine and hypotaurine. In addition, high elevations of the GABA content in the brain and in the CNS are not directly related to maximum anticonvulsive activities. Metabolic connections between GABA and other amino acids suggest that elevations in GABA can lead to changes in other amino acids. Pitkanen et al. (35) studied the levels of a number of amino acids, including GABA, taurine, and glycine in epileptic patients treated with vigabatrin. They found that the GABAergic neurotransmission is particularly affected by vigabatrin, but levels of glycine are also elevated.

Berthelot et al. (37) describe the synthesis of another GABA agonist, baclofen (-p-chlorophenyl-GABA), which binds to the GABA$_b$ receptor. Indeed, while many GABA a binding agonists have been investigated, few compounds that bind to GABA$_b$ were subjected to structure–activity studies. Berthelot et al. proved the specificity of baclofen analogues for the bicculine-insensitive GABA$_b$ receptor.

Another problem related to GABA is anoxia/ischemia in the central nervous system. An autoprotective system against the injuries produced by this condition is found in the white matter (39), which contains functional GABA$_b$ receptors that respond to an anoxic flux of GABA, producing a receptor-mediated cascade involving protein kinase. This increases the resitance of the organism to anoxia.

The GABA$_a$ and GABA$_b$ receptors also play a role in Ca2+ homeostasis and transmitter release in cerebellar neurons (40). It has been found that the receptors are coupled to the regulation of transmitter release, as well as intracellular Ca2+ homeostasis. The release elicited by moderate depolarization can be inhibited by either of the two types of receptor agonists, while if the depolarization is stronger, only GABA$_b$ agonists can inhibit it. The GABA$_b$ receptors are also more clearly involved in the regulation of the Ca2+ influx than the GABA$_a$ receptors.

Moss et al. (41) studied the regulation of GABA$_a$ receptors by multiple protein kinases, using a combination of molecular and physiological approaches. The results suggest that the GABA$_a$ receptors are under the control of multiple-cell signaling pathways, which can have an effect on inhibitory synaptic transmission mediated by GABA$_a$ receptors.

The GABA$_b$ receptor was studied from the point of view of its functional analysis by Hirouchi et al. (42) using a reconstituted system with purified GABA$_b$ receptor, Gi/Go protein, and adenyl cyclase. It was concluded that there is multiplicity in the GABA$_b$ receptors. It also was demonstrated that the inhibition of GABA$_b$ receptor binding by ethanol was accompanied by a decrease in the forskolin inhibition and a stimulated cAMP accumulation.

An important molecule with which GABA interacts is L-Dopa. Misu et al. (43) showed that depressor responses to L-Dopa in the rat nucleus tractus solitarii were inhibited by GABA. Pressor responses to GABA were reduced by a competitive L-Dopa antagonist, which also increased GABA release. D-Dopa does not show the same effect.

As mentioned before, GABA is related to epilepsy. A variety of animal models were developed in order to study this disease, and seizures are induced by sensory stimulation in genetically predisposed animals (44). The point of these studies is to develop new highly specific antiepileptic agents.

When seizures in gerbils (45) were divided into "minor" and "major," it was found that some drugs such as ethosuximide and valproic acid are active against absences and myclonus epilepsies, while phenobarbital, phenytoin, and carbamazepine are active against the grand mal (generalized tonic clonic seizures).

In general, researchers have tried to develop drugs with antiepileptic activity by designing drugs capable of selectively activating the central GABA system. Indeed, GABA mimetics have shown anticonvulsant properties in various seizure models (46).

Another role played by GABA in living organisms is related to the chemical transmission in the retina, which is utilized by some acrinic cells (47–48). In nonmammalian vertebrates it has been suggested (49–50) that bipolar cells or interplexiform cells in the retina are GABAergic. Moran et al. (51) suggest that the release of GABA can be stimulated directly or indirectly by activating the depolarizing Glu/Asp receptors within the visual pathway of the retina. One of their previous studies (52) showed that depolarizing Glu/Asp receptor agonists stimulated massive release of GABA from chicks. Using biochemical and autoradiographic analyses, they performed (51) studies in order to assess the effects of aspartate, glutamate, and kainate on GABA release from rat, chick, and frog retinas. H3-labeled GABA was used for the studies. In frogs, it was found that some nuclei located in the outermost and in the middle parts of the inner nuclear layer appeared highly labeled. Both inner and outer plexiform layers also seemed to have accumulated radioactive GABA. Addition of glutamate, aspartate, or kainic acid caused large increases in the release rate of GABA.

In the chick retina, the label is concentrated by cells in the horizontal, amacrine, and ganglion cell layers, with the horizontal layer being the most labeled (51). As in frogs, the aspartic, glutamic, and kainic acids stimulate strongly the release of GABA. Neither in frogs nor in chicks is there accumulation of GABA in the Müller cells. Conversely, the rat retina shows accumulation of GABA in the Müller cells and not in the horizontal cells. In the rat, aspartic, glutamic, and kainic acids do not stimulate GABA release.

In order to understand better the various biological aspects of GABA, its structure has to be investigated. A number of studies applied quantum-chemical methods to the determination of the geometry and order of stability of different conformers of GABA.

Fugler-Domenico et al. (53) used the Gaussian program with the 6-31G basis set to investigate the optimum geometries and the energies of some comformers of GABA and of aminooxyacetic acid (AOAA). The interest in the latter was due to its ability to cross the blood–brain barrier, an ability lacking in GABA, and to the fact that AOAA is a potent inhibitor of GABA-transaminase (54–55). Three conformers of GABA were studied: an extended one, a partially folded one, and a cyclic conformation that features a hydrogen bond between the hydrogen of the COOH group and the nitrogen of the amino group. The extended conformer of AOAA was found to be high in energy via preliminary optimization and therefore was not investigated further. As such, three conformers of AOAA were geometry optimized: a partially folded one, a cyclic one that features a hydrogen bond between the hydrogen of the COOH group and the nitrogen, and a cyclic conformer that features a hydrogen bond between the same hydrogen, but with the oxygen of the oxy group. In addition, this study investigated the conformations of GABA imine. GABA imine also exhibits a high-energy extended conformation, so only partially folded and cyclic conformers were investigated.

The starting geometry for all the species was neutral, not the zwitterion, and therefore the proton was positioned on the carboxyl's oxygen, not on the amino's nitrogen. Indeed, the large difference between the proton affinities of the COO- group and of the NH_2 group (which is much smaller) precludes the formation of the zwitterion in gas phase.

All the calculations performed by Fugler-Domenico et al. (53) were in gas phase, which, as a model for nonpolar solvents, is relevant in this case. Indeed, the issue here is the solvation in lipids, necessary for crossing the blood–brain barrier.

Out of the investigated conformers (which do not necessarily represent global minima, due to geometrical restrictions), the most stable for both GABA and GABA imine was found to be the partially folded conformation that does not feature a hydrogen bond of the OH . . . N type. The extended structure for GABA proves to be higher in energy than the partially folded one by 1.82 kcal/mole. The hydrogen-bonded cyclic structure of GABA is higher by 4.96 kcal/mole than the partially folded one. In the case of GABA imine, the cyclic one is higher in energy by 3.07 kcal/mole than the partially folded one.

In the case of AOAA, the cyclic conformer, with a hydrogen bond, is the most stable. The partially folded structure is higher by 3.20 kcal/mole, and the extended one is higher in energy than the cyclic one by 4.39 kcal/mole.

It was thus assumed by Fugler-Domenico et al. (53) that cyclic structures that might be less polar than an extended one are more lipophilic, and, as such, more capable of crossing the blood–brain barrier. An argument in favor of this hypothesis is that 5-fluoro-4-amino-pentanoic acid, a potent inhibitor of GABA transaminase in which there is a hydrogen bond between the fluorine and the O–H, also crosses the blood–brain barrier. γ-Keto acids are also found to cross it. In consequence, the ability to cross the blood–brain barrier might be related to the preference for a cyclic conformation.

A more recent study of GABA using ab initio calculations was performed by Ramek and Flock (56). The authors point out that although the neutral species is more stable in gas phase, in crystal phase the zwitterions are more stable. To study the neutral species with experimental methods is rather hard: Most amino acids decompose before melting. Therefore, quantum-chemical calculations provide a convenient tool for the investigation of various conformers of these molecules.

Ramek and Flock (56) apply ab initio calculations, at Hartree–Fock level, using the 4-31G basis set to the study of the potential energy surface of GABA. They use the Gamess (57) program, verifying that for each minimum the eigenvalues of the Hessian matrix are positive.

The study found a total of 122 local minima on the potenatial energy surface, with two of them of Cs symmetry and the others of C1 symmetry, forming 60 pairs of mirror images. The isomers that feature a positive value of the dihedral C–C–C–N angle are labeled I, II, etc. while the isomers featuring a negative value of this angle are labeled Im, IIm, etc. (56).

The optimization of the geometries was carried out without restrictions, as opposed to (53), where restrictions were imposed. However, both studies agree on the fact that the global minimum is a partially folded conformer, not featuring an O–H . . . N hydrogen bond. In (53), the dihedral angles found at 6-31G level were 61.6° for the C–C–C–N angle, 74.7° for the C–C–C–C angle, and –61.5° for the O–C–C–C angle, where the O represents the carbonyl oxygen. The same angles optimized without restrictions in (56) take values of 51.8°, –80.4°, and –27.9°.

Ramek and Flock found two sets of conformers with the O–H . . . N hydrogen bond: a low-energy set, with relative energies less than 16 kJ/mole, and a high-energy set, with relative energies higher than 30 kJ/mole.

These two sets, as pointed out by Ramek and Flock, correlate almost perfectly with the orientation of the -COOH group: The conformers with syn periplanar orientation of the CO and OH groups are part of the low-energy set, while the conformers with antiperiplanar orientation of these groups (with the exception of two conformers) are part of the higher-energy set.

The two conformers that form an exception by belonging to the low-energy set while featuring an antiperiplanar orientation exhibit a cyclic arrangement of the N–C–C–C–CO–O–H atoms in which N and the hydrogen of the OH form an intramolecular hydrogen bond. Thus, a seven-membered ring is obtained, which contains the hydrogen bond. In the first of the two conformers, the ring features the nitrogen and three carbons in almost the same plane, with the fourth carbon sticking out of the plane. The other conformer features two of the carbons sticking out of the plane formed by the other atoms, forming a boat shape. The distances between the nitrogen and the H of OH are 1.765 A and 1.884 A, respectively, typical lengths of hydrogen bonds (56). Rotations of the OH and COOH groups in reaction paths where the reactants are nonhydrogen-bonded conformers can lead to the formation of conformers with the OH . . . N hydrogen bond. In addition, there are reactions between conformers where the hydrogen bond is preserved.

Besides the intramolecular O–H . . . N hydrogen bond, Ramek and Flock point out the existence of other intramolecular interactions (56). The stronger of these is of electrostatic origin and occurs in the syn-periplanar orientation of the C=O and OH groups. The stabilization due to this interaction is estimated to be about 34 kJ/mole. In addition, there are attractive interactions of the type N–H . . . O=C, N–H . . . OH, and others present in every GABA conformer.

These interactions do not influence bond lengths, but some influence vibrational frequencies.

When GABA was conpared to glycine and beta-alanine, Ramek and Flock found that the GABA conformers are much more flexible than those of the other two amino acids. They suggest that this parallels influence of the ring size on the stability of the hydrogen bond, which is less constrained in GABA.

When the reactions of the transformation of one conformer into another are examined, Ramek and Flock (56) show that the preservation of the hydrogen bond leads to potential barriers of 18.68 kJ/mole and 15.21 kJ/mole. Reactions that do not preserve the hydrogen bond feature higher potential barriers of about 24 kJ/mole. They also state that only one conformer shows hydrogen bond character for the N–H . . . O=C interaction. In all the other conformers this interaction is purely electrostatic (56).

The structures described in (53), which are subjected to restrictions, and, as such, are not real minima but only low-energy configurations, are found to be transition states for some of the interconformer transformations (56).

In order to gather information about the way GABA functions at the molecular level within the synapse, a number of molecules containing a GABA-like fragment have been synthesized and tested for their agonistic activity in vivo and in vitro (57). To complement these studies, Lipkowitz et al. (58) performed an experimental and theoretical study for the elucidation of some of these compounds' structures. They applied X-ray crystallography, quantum mechanics, and molecular graphics to the comparison of GABA's structure to that of its antagonists.

The molecules investigated are muscimol, isoguvacine, trans-3-aminocyclopentane carboxylic acid, and 4,5,6,7-tetrahydroisoxazolo (5,4C)-pyridin-3-ol (THIP) (59). These molecules were chosen by Lipkowitz et al. (58) not only because they are potent agonists of GABA, but also for their structural diversity and because data were available for their inhibitory potency (IC50) agaist GABA (60) from the same research group, ensuring thus the same temperature GABA concentration and cerebral tissue concentration during the experiment.

The X-ray analysis was performed by the authors on Nicolet R3m automatic diffractometers. The quantum calculations were performed with semiempirical methods (MNDO and AM1) as implemented by the MOPAC computer program (61). The molecules were assumed to be singlets, and the geometrical parameters were fully optimized.

The quantum-chemical calculations performed by Lipkowitz et al. describe atomic charges on the atoms of the four molecules investigated. They also describe the charge distribution in GABA and its agonists by defining a cationic charge as the charge of the nitrogen and four attached carbons, and the anionic charges as the sum of charges on the carboxylate moiety and the atoms close to it, as relevant in the different molecules. They compute the charge separation as the distance between the nitrogen of the amino group and the carbon of the carboxylate (in the case of muscimol and THIP, the isoxazole group plays the role of the carboxylate). No correlations were observed between the IC50 data and the distance between charges, with the differences in charges, or with the molecular dipole. It was inferred that the molecular recognition process is more complicated than one based on simple structural and electrostatic effects.

A factor that may influence the value of IC50 is that the agonists leave the bulk aqueous phase and might be attracted in a hydrophilic or hydrophobic way to the receptor. If this were an important factor, a linear correlation would be obtained between percent polar surface area and IC50. However, experiments (58) show this correlation to be nonlinear, suggesting that other effects are involved in the agonist–receptor binding. When redox reactions were investigated (58), it was also found that they alone are not responsible for substrate–receptor interaction. It might be concluded that a combination of all these factors accounts for the binding.

Lipkowitz et al. (58) show that GABA and its agonists can adopt conformations where the carboxylate, or the group that plays its role, is superimposable and at the same time the ammonium centers are superimposable, resulting in a single "active conformation." However, the cost in energy to adopt this conformation, as calculated by AM1 and MNDO semiempirical methods by Lipkowitz et al. (58), is quite high. Muscimol can be superimposed on GABA with minor conformational changes, but it cannot be superimposed as well on the other agonists of GABA. The authors conclude that the strong binding of muscimol to the GABA receptor is due to the position of its ammonium group, which lies off the carboxylate vector and is closer to negative charges on the receptor. GABA, being a flexible molecule, can also adopt this conformation, while the other agonist cannot, a fact that makes it less effective.

The fact that GABA has been studied so far only with double-zeta basis sets, without the inclusion of polarization functions and correlation energy effects, made it desirable to perform calculations on some of the GABA conformers at a higher calculational level than previously.

Sapse and Jain (62) have investigated three conformers of GABA as shown in Figure 3.1.

The basis set used is the 6-31G** basis set, which is a double-zeta set, including d polarization functions on the nonhydrogen atoms and p

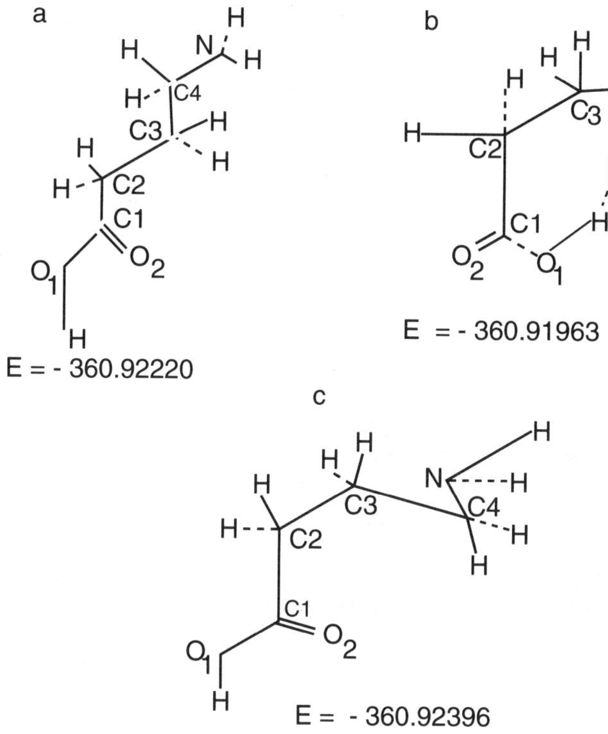

FIGURE 3.1 a. Extended conformer of GABA. b. Cyclic conformer of GABA. c. Partially folded conformer of GABA.

polarization functions on the hydrogens. Using the Gaussian-92 computer program with total energy optimization, it was found that the energy of the conformer 3.1a is −360.92220 au at Hartree–Fock level; conformer 3.1b features an energy of −360.91963 au, and conformer 3.1c an energy of −360.92396 au, also at Hartree–Fock level. Therefore, setting the relative energy of the most stable conformer, 3.1c, at 0.0, the relative energies of the other two are 1.1 kcal/mole for 3.1a and 2.7 kcal/mole for 3.1b. It can thus be seen that the order of stability is the same as for the three conformers investigated in (53), which are similar in geometry. However, that work uses the 6-31G basis set, without polarization functions, which leads to an increase of the relative energies.

When the correlation energy is estimated via MP2/6-31G** calculations, using the Hartree–Fock 6-31G**-btained geometry, the following energies are obtained: 3.1a −362.00505, 3.1b −362.00795, and 3.1c −362.00822.

It can be seen that using the correlation energy effects, the stabilities of a and b are reversed, while c remains the most stable. The relative energies of a and b are now 2 kcal/mole and 0.2 kcal/mole, respectively. It is thus hard to state that there is difference in stability between b and c, leading us to believe that the correlation energy stabilizes the hydrogen-bonded conformer b.

It may be concluded that in order to obtain an accurate description of the potential energy surface, it is necessary to use basis sets including polarization functions, as well as correlation energy calculations.

References

1. Schade, J.P., and Ford, D.H. *Basic Neurology*. Elsevier, New York, 1973.
2. McGeer, P.L., Eccles, J.C., and McGeer, E.G. *Molecular Neurobiology of the Mammalian Brain*. Plenum Press, New York, 1978.
3. Cooper, J.B., Bloom, F.E., and Roth, R.H. *The Biochemical Basics of Neuropharmacology*. Oxford University Press, New York, 1982.
4. Turner, A.J., and Whittle, S.R. *Bichemistry 209*, 29, 1983.
5. Zepeda, A.T., Ortiz, F.J., Mendez-Franco, J., Otero-Siliceo, E., and Perez dela Mora, M. *Amino Acids 9*, 3, 207, 1995.
6. Krogsgaard-Larsen, P., and Christiansen, T.R. *Eur. J. Med. Chem. 14*, 157, 1979.
7. Krogsgaard-Larsen, P., Falch, E., and Jacobsen, P. In *Actions and Interactions of GABA and Benzodiazepines*. Raven Press, New York, 1984.
8. Falch, E., Hedegaard, A., Nielsen, L., Jensen, B.R., Hjeds, H., and Krogsgaard-Larsen, P. *J. Neurochem. 47*, 3, 898, 1986.
9. Zorn, S.H., Willmore, L.J., Bailey, C.M., and Enna, S.J. In *Neurotransmitters, Seizures and Epilepsy III*. Raven Press, New York, 1986.
10. Krogsgaard-Larsen, P., Falch, E., Scousboe, A., and Curtis, D.R. In Neurotransmitters, Seizures and Epilepsy III. Raven Press, New York, 1986.
11. Burke, D., Andrews, C.J., and Knowles, L. *J. Neurol. Sci. 14*, 199, 1971.
12. Santi, M.R., Cox, D.H., and Guidotti, A. *J. Neurochem. 50*, 4, 1080, 1988.
13. Haring, P., Stahli, C., Schoch, P., Takacs, B., Staehelin, T., and Mohler, H. *Proc. Natl. Acad. Sci. USA 82*, 4837, 1985.
14. Leeb-Lundberg, L.M.F., and Olsen, R.W. *Mol. Pharmac. 23*, 315, 1983.
15. Ahton, D., Geerts, R., Waterkeyn, C., and Leysen, J.E. *Life Sci. 29*, 2631, 1981.
16. Mandel, P., and DeFeudis, F.V. *Advances in Experimental Medicine and Biology: GABA Biochemistry and CNS Function*. Plenum Press, New York, 123, 1979.
17. Krnjevic, K., and Schwartz, S. *Exp. Brain. Res. 3*, 320, 1967.
18. Silverman, R.B., Durkee, S.C., and Invergo, B.J. *J. Med. Chem. 29*, 764, 1986.
19. Johnston, G.A.R., Curtis, D.R., Beart, P.M., Game, C.J.A., McCullough, R.M., and Twitchin, B.J. *J. Neurochem. 24*, 157, 1975.
20. Bergmann, K.J., *Clinical Neuropharm. 8*, 1, 13, 1985.
21. Scatton, B., Zivkovic, B., and Dedek, J. *J. Pharmacol. Exp. Ter. 220*, 689 1982.
22. Mackay, A.V.P., Davies, P., Dewar, R.J., and Yates, C.M. *J. Neurochem. 30*, 827, 1978.
23. Emson, P.C., and Lindvall, O. *Neuroscience. 4*, 1 1979.
24. Ribak, C.E. *J. Neurocyt. 7*, 461, 1978.

25. Ribero da Silva, A., and Coimbra, A. *Brain Res. 188*, 449, 1980.
26. Goldensohn, E.S., and Purpura, D.P. *Science 139*, 1963.
27. Matsumoto, H., and Ajmone-Marsan, C. *Exp. Neurol. 9*, 286, 1964.
28. Lott, I.T., Coulombe, T., DiPaolo, R.V., Richardson, E.P., and Levy, H.L. *Neurology 28*, 47, 1978.
29. Lloyd, K.G., Munari, C., Worms, P., Bossi, L., and Morselli, P.L. In *Epilepsy an Update on Research and Therapy*, A.R. Liss, New York, 1983.
30. Barker, J.L., and Nicole, R.A. *Science 176*, 1043, 1972.
31. Davidoff, R.A. *Science 175*, 331, 1972.
32. Hornykiewicz, O., Lloyd, K.G., and Davidson, L. In *GABA in Nervous System Function*, Raven Press, New York, 1976.
33. Manyam, B.V. *Arch. Neurol. 39*, 391, 1982.
34. Teychenne, P.F., Ziegler, M.G., Lake, C.R., and Enna, S.J. *Ann. Neurol. 11*, 76, 1982.
35. Pitkanen, A., Matilainen, R., Ruutiainen, T., Lehtinen, M., and Riekkinen, P. *J. Neurology, Neurosurg., and Psy. 51*, 1395, 1988.
36. Grove, J., Palfreyman, M.G., and Schechter, P.J. *Clin. Neuropharmacol. 6*, 223, 1983.
37. Berthelot, P., Vaccher, C., Musadad, A., Flouquet, N., Debaert, M., and Luyckx, M. *J. Med. Chem. 30*, 743, 1987.
38. Schlewe, G., Wermuth, C.G., and Chambon, J.P. *Eur. J. Med. Chem.-Chim. Ther. 19*, 2, 181, 1984.
39. Fern, R., Ransom, B.R., and Waxman, S.G. *Mol. Chem. Neuropathol. 27*, 2, 107 1986.
40. Schousboe, A., Elster, L., Damgaard, I., Pallos, H., Krogsgaard-Larsen, P., and Kardos, J. *GABA: Recept. Transp. Metab. 103*, 1996.
41. Moss, S.J., McDonald, B., Gorrie, G.H., Krishek, B.K., and Smart, T.G. *GABA: Recept. Transp. Metab. 173*, 1996.
42. Hirouchi, M., Mizutani, H., Nishikawa, M., Nakayasu, H., and Kuriyama, K. *GABA: Recept. Transp. Metab. 227*, 1996.
43. Misu, Y., Yue, J.L., Okumura, Y., Miyamae, T., and Goshima, Y. *GABA: Recept. Trans. Metab. 115*, 1996.
44. Loscher, W., Frey, H.H., Reiche, R., and Schultz, D. *J. of Pharmac. and Exp. The. 226*, 3, 839, 1983.
45. Frey, H.H., Loscher, W., Reiche, R., and Schultz, D. *Neuropharmacology 20*, 769, 1981.
46. Worms, P., and Lloyd, K.G. In *Neurotransmitters, Seizures and Epilepsy*. Raven Press, New York, 1981.
47. Brecha, N. In *Chemical Neuroanatomy*. Raven Press, New York, 1985.
48. Neal, M.J. *General Pharmac. 7*, 321, 1976.
49. Wu, S.M., and Dowling, J.C. *Proc. Natl. Acad. Sci. USA 75*, 5205, 1978.
50. Yazulla, S., and Brecha, N. *Invest. Ophtal. 19*, 1415, 1980.
51. Moran, J., Passantes-Morales, H., and Redburn, D.A. *Brain Res. 398*, 276, 1986.
52. Moran, J., and Passante-Morales, H. *J. Neur. Sci. Res. 10*, 261, 1983.
53. Fugler-Domenico, L., Russell, C.S., and Sapse, A.M. *J. Mol. Structure (THEOCHEM) 168*, 323, 1988.
54. Nishibori, M., Oishi, R., Itoh, Y., and Saeki, K. *Japan. J. Pharmacol. 41*, 403, 1986.
55. Krogsgaard-Larsen, P. *J. Med. Chem. 24*, 1377, 1981.

56. Ramek, M., and Flock, M. *Amino Acids 8*, 271, 1995.
57. Schmidt, M.W. et al. GAMESS *QCPE Bulletin 10*, 52, 1990.
58. Lipkowitz, K.B., Gilardi, R.D., and Aprison, M.H. *J. Mol. Structure 195*, 65, 1989.
59. Krogsgaard-Larsen, P., and Johnson, G.A.R. *J. Neurochem. 30*, 1377, 1978.
60. Johnson, G.A.R., Allan, R.D., Kennedy, S.M.E., and Twitchen, B. In *GABA-Neurotransmitters*. Munksgaard, Copenhagen, 1979.
61. Stewart, J.P. *QCPE Bull. 3*, 2, 455, 1983.
62. Sapse, A.M., and Jain, D.C. Unpublished results.

4
The Diaminobutyric (DABA), Delta Aminopentanoic, and Epsilon Aminohexanoic Acids

The L isomer of DABA, 2,4-diaminobutyric acid, is a major component of the polymixin group of antibiotics and a component of bacterial cell walls and of certain Lathyrus and related seeds (1–2). It was found to be neurotoxic in rats and mice (3), though it may not be neurotoxic in humans. L-DABA is widely distributed in nature (4).

The termination of neurotransmission mediated by gamma aminobutyric acid (GABA), which, as shown before, is the main inhibitory amino acid neurotransmitter in the mammalian central nervous system, takes place via its reuptake through a sodium-dependent carrier protein (5).

DABA, 2,4-diaminobutyric acid, which features a structure similar to that of GABA, with an additional amino group in the alpha position, has been reported to inhibit GABA uptake competitively (6–7). It is also thought that DABA stimulates the release of GABA, probably via a 1:1 hetero-exchange (8). Experimental observations have led to the conclusion that DABA is transported by the same protein as GABA (7).

Erecinska et al. (5) have shown that DABA is taken up by synaptosomes, through a high-affinity system inhibited by GABA. The uptake depends on the second power of the transmembrane electrical potential and on the first power of the sodium concentration gradient. When kinetics studies were undertaken (5), it was found that the transport of H3 radioactive DABA was saturable and followed Michaelis–Menten kinetics. The uptake of radioactive DABA was inhibited competitively by GABA. On the other hand, measuring the uptake of radioactive GABA with and without DABA showed also that DABA inhibits the GABA uptake in a competitive manner.

It was found (5) that the Michaelis constant for DABA uptake is pH independent in the 6.5–7.8 range, which indicates that $DABA^+$ binds to the carrier. Indeed, the $DABA^+$ concentration does not change much in this pH range. Since DABA metabolizes slowly, it is convenient to use in transport studies, using radioactive isotopes, whose distribution reflects the DABA and not its metabolites.

Erecinska et al. (5) conclude that sodium ions are cotransported with the amino acid from outside to inside. In contrast with GABA, which shows sigmoid properties related to the sodium concentration, DABA uptake shows linear dependence on the concentration of external sodium. According to Radian and Kanner (9), the stoichiometry for the translocation cycle catalyzed by the GABA transporter in synaptic plasma membrane vesicles is composed of two sodium ions and one chlorine ion for each GABA zwitterion. However, Erecinska et al. (5) state that the transport of GABA can be inhibited or stimulated by ions that are not cotransported with it, and secondary movements of anions might be necessary in order to ensure neutrality while sustaining amino acid uptake. The findings of the study imply that the fully loaded GABA carrier has a charge of 2+, while the DABA transport, because it requires only one sodium ion per DABA ion, utilizes only half the energy available in the sodium ion concentration gradient. Thus, the concentrations of DABA are lower than those of GABA. The explanation of why DABA requires only one sodium ion instead of two as in GABA might be related to the fact that the second amino group, which is positively charged at neutral pH, takes the place of one of the two sodium ions on the carrier protein, indicating that one of the sodium binding sites is close to the amino acid binding site.

As mentioned before, DABA is found to be toxic in mice and rats. Chen et al. (4) suggest, on the basis that L-DABA inhibits ornithine carbamoyltransferase, that this toxicity is partly due to an ammonia toxicity, but the major cause of toxicity is the penetration of L-DABA into the brain. Indeed, the observed toxic symptoms do not depend on changes in the concentrations of any other amino acids in the brain. It is true, though, that as the toxicity persists, GABA concentrations increase in the brain. Chen et al. report great increases of L-DABA concentration in rats' brains following intraperitoneal injections with L-DABA. The concentration of other amino acids, such as glutamine, asparagine, and GABA, increase slightly. However, it was found (10) that while L-DABA injections produce tremors and convulsions, D-DABA does not cause toxic symptoms. This difference might be due to the fact that, as shown by the analysis of DABA content of the brain following intraperitoneal injections with D-DABA, D-DABA does not penetrate the blood–brain barrier. Accordingly, the presence of DABA in the brain is necessary for toxicity to occur.

It was found also that DABA has some activity against at least two types of seizures (11–12). Rostain et al. (13) studied the effect of environmental pressure on rats in order to investigate the high-pressure nervous syndrome. The rats injected with DABA and subjected to pressure died during decompression, and it was therefore not possible to determine whether the damage was due to pressure and decompression or to the treatment with

DABA. However, it was found that DABA and nipecotic acid prevented the increase in theta and/or delta waves in the brain.

Meldrum et al. (14) showed DABA to be a neuron-selective transport inhibitor that functions as a convulsant or proconvulsant after intracerebroventricular injections.

In order to understand better the biological activity of DABA, especially the ability of the L isomer to penetrate the blood–brain barrier, information had to be obtained about its structure. Hinazumi and Mitsui (15) have determined the crystal structure of L-DABA by X-ray diffraction. They compare its bond lengths and angles to those of lysine and ornithine.

Single crystals of L-DABA, grown from aqueous solution at about 5°C, were colorless plates elongated along the b axis. The group space was suggested to be P2 or P2/m, and the first of these was adopted because the compound was optically active. The unit call parameter was obtained from a least-squares procedure in which the differences between the sine theta square observed and sine theta square calculated were examined.

The results indicate for the C_1C_2 bond length (see Figure 4.1) a value of 1.538 A, for the C_2C_3 bond length a value of 1.536 A, and for the C_3C_4 a value of 1.515 A. The N1C2 and N2C4 feature bond lengths of 1.491 A and 1.499 A, respectively. The structure of the molecule is extended.

Comparing the values to lysine and ornithine, the authors conclude that the bond angles and bond lengths in DABA are quite similar to those observed in the other two amino acids.

A theoretical study using ab initio quantum-chemical calculations was performed in addition to crystallographic studies to elucidate the preferred conformation of DABA and to obtain numerical values for its geometrical parameters. Sapse et al. (16) have performed such calculations, with the goal of providing insight into the similarities and differences in the transport behavior of GABA and DABA.

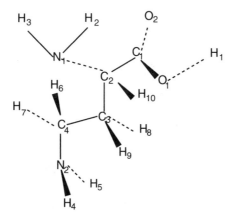

FIGURE 4.1 Extended conformer of DABA.

Ab initio calculations, as mentioned before, are performed for the gas phase, unless solvent effects are examined, but the gas phase may serve as a good model for the nonhydrated environment of the inner membrane leaflet, which must be traversed in the course of crossing the blood–brain barrier or the hydrophobic receptor-binding site.

The Gaussian program was used to perform ab initio self-consistent field calculations at Hartree–Fock level. Three types of conformations were considered: extended, partially folded, and cyclic. The cyclic conformations featured a hydrogen bond between the proton set on one of the carboxyl's oxygens and the gamma amino group (as a neutral species or as a zwitterion with this proton set on the nitrogen). In addition, the cations of DABA, featuring an additional hydrogen set on the gamma amino group, in cyclic conformations featuring a hydrogen bond either between this proton and the nitrogen of the alpha amino group or between the same proton and the nonhydrogen-bearing oxygen of the carboxyl, were also investigated.

These conformations were subjected to geometric optimization, using the Berny optimization method (17). The basis set used was the 6-31G basis set.

Figures 4.1–4.4 show the conformations investigated. The extended conformation shown in Figure 4.1 was geometry-optimized, keeping the N2–C4–C3–C2 angle and the C4–C3–C2–C1 angle at 180.0°. Several folded conformations were tried, and preliminary optimizations were performed. It was found that the conformation featuring the C4–C3–C2–C1 angle of 90.0° was found to be the most stable and was singled out for further optimization. This structure is shown in Figure 4.2. A structure featuring a hydrogen bond between H1 and N1 was discarded as too high in energy. Structure 3, shown in Figure 4.3, which features a hydrogen bond between H1 and N2, was geometry-optimized, starting from an initial geometry that ensures the formation of the bond. During the optimization the N2–H1–O1–C1 angle was kept at 0.0°.

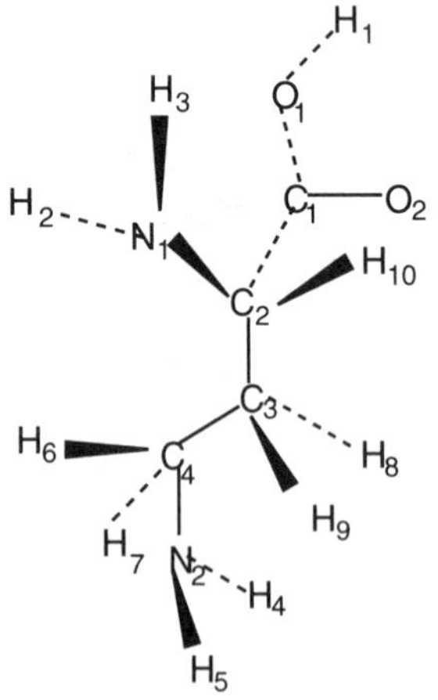

FIGURE 4.2 Partially folded conformer of DABA.

Figure 4.4 depicts the cationic forms of DABA, Figure 4.4a featuring the protonation on N1, 4.4b with an additional proton on N2 hydrogen-bonded to N1, and 4.4c with the same proton hydrogen-bonded to O2.

It is found that structures 4.1, 4.2, and 4.3 are almost equal in energy. That of 4.1 is the most stable, followed by 4.3, higher in energy by

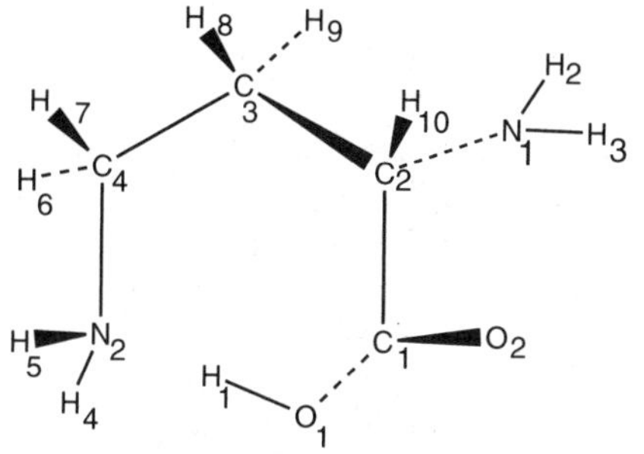

FIGURE 4.3 Cyclic conformer of DABA.

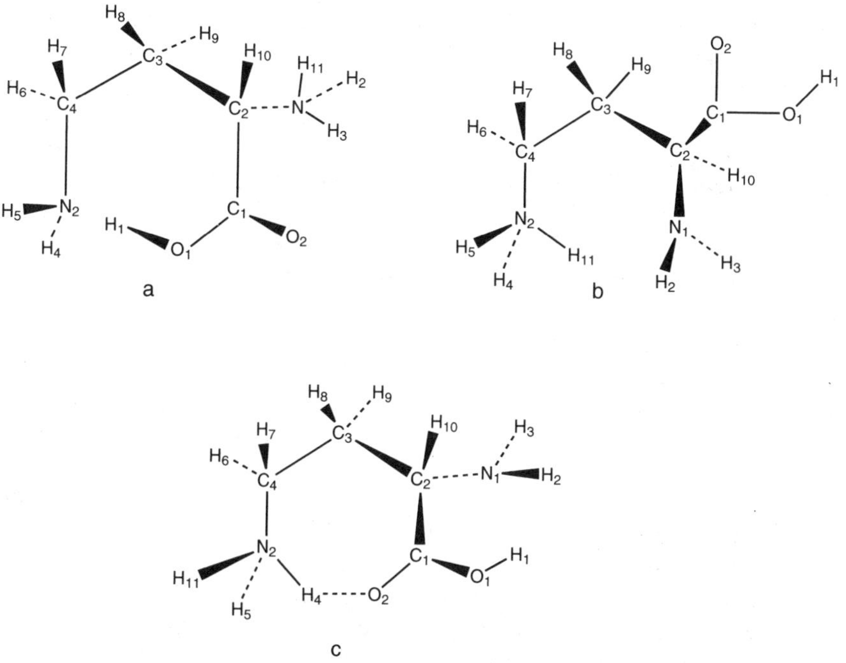

FIGURE 4.4 a. Cyclic conformer of the DABA cation with H at N_1. b. Cyclic conformer of the DABA cation with H at N_2. c. Cyclic conformer of the DABA cation with H at N_2 and featuring a hydrogen bond between N_2 and O_2.

0.5 kcal/mole, and structure 4.2 is higher in energy than structure 4.1 by 0.6 kcal/mole.

The hydrogen bond present in structure 4.3 features a H1–N2 distance of 1.7 A. When a number of other partially folded structures were examined, it was found that structure 4.2 is indeed the most stable of partially folded conformations.

It is possible that in a hydrated medium, the zwitterion might be more stable than the neutral species. However, in gas phase, a cyclic zwitterion was found to be higher in energy by 25 kcal/mole than its neutral equivalent.

The net atomic charges were examined, using Mulliken population analysis. It was found that C1 features more of a positive charge in the 4.3 conformation than in the other conformations investigated, with O1 bearing a larger negative charge. It was also found that the charge separation between C1, O1, and H1 is higher in the cyclic form of DABA than in the cyclic form of GABA. This might contribute to the stability of the 4.3 conformation.

In addition to total energies of the neutral molecules, the study calculated their proton affinities. Each conformation was protonated either at N1 or at N2, the energy of each cation thus formed was calculated, and the energy of the neutral species was subtracted from it. The result, in atomic units, was multiplied by a factor of 627.5, in order to obtain the protonation energy in kcal/mole. To obtain the proton affinity of the carboxyl group, the energy of the anion obtained by removing the proton on the carboxyl's oxygen was calculated and subtracted from the energy of the corresponding neutral species. The proton affinities thus obtained are 354.6 kcal/mole for the protonation of the anion corresponding to structure 1, 355.1 for structure 2, and 354.5 for structure 3. The amino groups' proton affinities range from 225.8 kcal/mole for structure 2 protonated at N1 to 242.0 kcal/mole for cyclic structures protonated at N2.

The proton affinities of the amino groups and the carboxyl group may be important factors bearing on the lipophilicity and the ability of L-DABA to cross the blood–brain barrier. A higher proton affinity at NH2 facilitates its protonation, while a low proton affinity at COO- leads to the formation of the anion. The most stable cation for L-DABA was found to be the conformation featuring a cycle, with the proton positioned on N2 and hydrogen-bonded either to N1 or to one of the carboxyl's oxygens. The structures are more stable than those featuring the proton on N1. When these results are compared to the GABA results (see Chapter 3), one sees that the highest proton affinity for the amino group of L-DABA, 242 kcal/mole, is higher than the corresponding value of 237.3 kcal/mole for GABA. On the other hand, the carboxylate's proton affinity for GABA, 362.6 kcal/mole, is higher than the one for DABA, namely 354 kcal/mole. These values seen to indicate that the formation of the zwitterion is more favorable for DABA than for GABA. Accordingly, it suggests a lower lipophilicity for DABA, which means that diffusion alone cannot account for its ability to cross the blood–brain barrier. As such, this process may involve mediated transport of DABA or diffusion by uncharged species in a cyclic conformation. Indeed, it seems that the presence of a cyclic conformation increases the ability of some amino acids to cross the blood–brain barrier (16).

Several experimental studies (18–19) have shown that L-DABA strongly inhibits high-affinity GABA uptake in synaptosomal tissue. Debler and Lajtha (20) showed that L-DABA inhibits 89% high-affinity GABA uptake in synaptosomal P2 tissue. It has been suggested that the synaptosomal binding sites for GABA are specific and recognize L-DABA. This might be due to the apparent flexibility of L-DABA (16). The small difference between the energies of different conformations of L-DABA makes it relatively easy for the molecule to adopt a conformation complementary to a GABA binding site. The flexibility of L-DABA might also account for its inhibition of taurine and glycine uptake (20).

The comparison between proton affinities of GABA and DABA is compatible with Erecinska's findings (5) that GABA is cotransported in synaptosomal tissue by a high-affinity carrier with two Na^+, while DABA is transported with only one Na^+. Indeed, at neutral pH, DABA is cationic and GABA is neutral. In conclusion, L-DABA was found to be a flexible molecule with very stable cations.

Delta-Aminopentanoic Acid

In a paper that studies glycine, beta-alanine GABA, and delta-aminopentanoic acid, Ramek (21) finds for delta-aminopentanoic acid a geometry resembling the chair form of cyclohexane, with the atoms of the COOH group, the alpha carbon, and the nitrogen forming one plane, and the beta, gamma, and delta carbons forming an approximately parallel plane. The method used was ab initio, with the 4-31G basis.

In a subsequent paper, delta-aminopentanoic acid, DAPA, was studied also with ab initio methods, from the point of view of the structure of H-bonded conformers. Ramek (22) compares the hydrogen bond in DAPA to those in its homologues and also to bimolecular adducts formed by HOOH, CH_3COOH, and C_2H_5COOH. In addition, he studies hydrogen-bond formation in NH_3, CH_3NH_2, and $C_2H_5NH_2$.

The Gamess program (23) was used to perform restricted Hartree–Fock calculations (RHF), with the 4-31G basis set. This basis set, which features two Slater orbitals for the valence electrons, one expanded in a series of three Gaussian functions and the other approximated by one Gaussian and one Slater orbital for the core electrons expanded in a series of four Gaussian functions, was found to be quite reliable for this type of compound (22), and numerous calculations on other amino acids, such as glycine, alanine, and GABA, are available for comparison.

A number of conformers and transition states were optimized, and their nature as a minimum or as a transition state was determined by calculating the vibration frequencies. The minima feature only real frequencies, while transition states feature one imaginary frequency. The four conformers reported exhibit a strong O–H...N hydrogen bond and energies of −399.34076 au, −399.33972 au, −399.33923 au, and −399.33795 au. There is thus a difference of at most 1.76 kcal/mole between the most and the least stable conformation. All four conformations feature fairly similar C3C4C5N dihedral angles (from 51.69° for the second conformation to 90.19° for the first). The C2C3C4C5 angle, though, is different, taking a value of −57.10° in the first conformation, 57.63° in the second, −80.49° in the third, and −134.59° in the fourth. The C1C2C3C4 angles and also different, taking values of −63.63°, −111.85°, 110.37°, and 69.28. The same is true for the O1C1C2C3 angle, with values of −96.44°, −140.93°, 113.24°, and −176.75°.

In order to undergo a transition from one conformation to another, the molecule features rotational barriers for those rotations around various bonds, which change the respective dihedral angles. For instance, if conformer I undergoes geometric changes that lead to its transformation into conformer II, the increase of the C2C3C4C5 angle and the decrease of the O1C1C2C3 angle go through a barrier of 22.7 kJ/mole. A barrier of 27.2 kJ/mole is found in the transition from I to IV, a transition that involves a decrease of the O1C1C2C3 angle, increase of the C1C2C3C4 angle, decrease of the C2C3C4C5 angle, and increase of the HNC5C4 angle. The changes conformer I has to undergo to transform into III require a barrier of 42.55 kJ/mole. The transformations of II into III or IV feature very similar barriers, of about 27 kJ/mole. In the reactions where one conformer is turned into another, the hydrogen bond is preserved.

As mentioned before, in addition to the calculations on delta-aminopentanoic acid, Ramek (22) also studied bimolecular adducts of the RCOOH...NH2R′ type. In these adducts, as in the amino acid, there is no proton transfer from the carboxyl to the amino group in vacuo. As is known, the proton affinity of the COO- anion is larger than that of the amino group, so these results are not surprising. In these calculations the CO and OH groups were kept in an antiperiplanar position, and R and R′ were set at a maximum distance from each other. As such, the energies thus obtained do not represent real minima, since the syn-periplanar position of the CO and OH groups are more stable, and there might be favorable C–H...O interactions. However, the point of the calculations was to optimize the arrangement of the O–H...N hydrogen bond, which is possible with CO and OH antiperiplanar (22).

When glycine, alanine, GABA, and DAPA are compared, the following trend is noticed (22): Glycine features only one conformer with the O–H...N bond, alanine two that are mirror symmetric to each other. GABA forms two such pairs, and DAPA forms four such pairs. It thus becomes clear that each additional CH_2 group doubles the number of conformers with the O–H . . . N hydrogen bond. Another observed trend is related to the number of reactions that preserve this hydrogen bond, which increases with the number of conformers, that is, it increases by one for each CH_2 group, per each conformer.

A comparison of the strength of hydrogen bonds with the number of atoms shows an increase with the ring size. There is a difference between DAPA and the other amino acids mentioned: The H-bonded conformers are not wells in the potential energy surface, as they are in glycine, alanine, or GABA. This difference might be due to the constraintless formation of the H-bonded conformers (22).

Epsilon-Aminohexanoic Acid

Epsilon-aminohexanoic acid has been studied with ab initio calculations by Ramek (23) as part of a study of the intramolecular hydrogen bonds and conformer stability of ω amino acids. It was found (23) that the potential surface of ε-aminohexanoic acid contains a total of fourteen symmetry unique local minima that feature an O–H . . . N hydrogen bond. The criterion used for symmetry uniqueness is the N–C–C–C torsion angle. The conformers featuring a positive value of this angle were labeled from I to XIV, while those featuring a negative value were labeled Im, IIm, etc. Contrary to the amino acids containing fewer than six carbon atoms, the reactions that preserve the O–H . . . N hydrogen bond in ε-aminohexanoic acid form a very complex pattern.

Two conformers, labeled VIII and XII and featuring values of the C–C–C–N torsion angle of 66.32° and 113.72°, respectively, exhibit a particularly small value for the lowest potential barrier. Indeed, in VIII, a combination of OH, COOH, and NH_2 rotations that breaks the O–H . . . N hydrogen bond leads to a change in energy of only 0.9 kJ/mole, with a vibrational frequency of 86.9 cm^{-1}. This corresponds to a zero-point energy of vibration of 0.52 kJ/mole. In XIII, the lowest potential barrier belongs to a reaction that preserves the O–H . . . N bond and is only 0.06 kJ/mole. The frequency associated with it is 40.2 cm^{-1}, with a zero-point energy of 0.24 kJ/mole. Accordingly, since the zero-point vibrational energy is larger than the potential barrier, XIII cannot be considered a stable isomer (23). The order of stability of the hydrogen bond decreases from I (C–C–C–N torsion angle of 71.29° to XIV (C–C–C–N torsion angle 31.19°). When the zero-point vibrational energy is included, the order is slightly changed, with XII (C–C–C–N torsion angle of 87.57°) more stable than higher-numbered conformers, right after IX (C–C–C–N torsion angle 128.71°). The kinetic stability, determined by the height of the potential barrier to reactions, shows conformer II (C–C–C–N torsion angle 53.2°) to be the most stable, with conformer XIII being the least stable.

Ramek (23) shows that the strength of the hydrogen bond increases with the number of carbons up to aminopentanoic acid. From the aminopentanoic acid to aminohexanoic acid it decreases. There are no significant changes in the energies of the most and least stable isomers among the amino acids with four, five, or six carbons. The ring formed in order to obtain the hydrogen bond O–H . . . N is the most stable in delta-aminopentanoic acid.

The number of atoms present in the delta aminoacid ring (eight) provides a minimum of constraint and as such is the most stable. In GABA the number of carbon atoms is too small to ensure a close contact between the H on oxygen and the nitrogen, while in epsilon-aminohexanoic acid there is too much distorted geometry, with a main torsion angle of 120.0°, characteristic of transition states. As opposed to amino acids with fewer carbons, the ε-aminohexanoic acid features conformers with nine-membered rings with the COOH group in a distorted syn orientation (23).

Extrapolating the results, Ramek (23) proposes that 10-membered rings will allow the formation of hydrogen bonds in conformers of amino acids or other organic compounds in which terminal groups exist in their most favorable orientation, and this could explain the presence of 10-membered rings in peptides and proteins.

References

1. Ressler, C., Restone, P.A., and Ehrenberg, R.H. *Science 134*, 188, 1961.
2. Bell, E.A., and Tirimana, A.S.L. *Biochem. J. 97*, 104, 1965.
3. Kessel, D. *Fed. Proc. Fed. Amer. Chem. Soc. Exp. Biol. 18*, 258, 1959.
4. Chen, C.H., Flory, W., and Koeppe, R.E. *Tox. and Appl. Pharmacology 23*, 334, 1972.
5. Erecinska, M., Troeger, M.B., and Alston, T.A. *J. Neurochem. 46*, 1452, 1986.
6. Iversen, L.L., and Johnston, G.A.R. *J. Neurochem. 18*, 1939, 1971.
7. Roskoski, R. *J. Neurochem. 36*, 1236, 1981.
8. Simon, J.R., Martin, D.L., and Kroll, M. *J. Neurochem. 23*, 981, 1974.
9. Radian, R., and Kanner, B.I. *Biochemistry 26*, 1236, 1983.
10. O'Neal, R.M., Chen, C.H., Reynolds, C.S., Meghal, S.K., and Koeppe, R.E. *Biochem. J. 106*, 699, 1968.
11. Horton, R.W., Collins, J.F., Anlezark, G.M., and Meldrum, B.S. *Eur. J. Pharmacol. 59*, 75, 1979.
12. Bichard, A.R., and Little, H.J. *Br. J. Pharmacol. 76*, 447, 1982.
13. Rostain, J.C., Wardley-Smith, B., Forni, C., and Halsey, M.J. *Neuropharmacol. 25*, 5, 545, 1986.
14. Meldrum, B.S., Croucher, M.J., and Krogsgaard-Larsen, P. *Excerpta Medica, 182*, 1982.
15. Hinazumi, H., and Mitsui, T. *Acta Cryst. B27*, 2152, 1971.
16. Fugler-Domenico, L., Russell, C.S., and Sapse, A.M. *Struct. Chem. 1*, 379, 1990.
17. Schlegel, H.B. *J. Comp. Chem. 3*, 214, 1982.
18. Iverson, L.L., and Kelly, J.S. *Biochem. Pharmacol. 24*, 933, 1975.
19. Krogsgaard-Larsen, P. *J. Med. Chem. 24*, 1377, 1981.

20. Debler, E.A., and Lajtha, A. *J. Neurochem. 48*, 1851, 1987.
21. Ramek, M. *Int. J. Quantum Chem. Quantum Biol. Symp. 17*, 45, 1990.
22. Ramek, M. *Structural Chem. 6*, 1, 15, 1995.
23. Ramek, M. *Int. J. Quantum Chem. Quantum Biol. Symp. 21*, 79, 1994.

5
Ab Initio Studies of Some Acids and Basic Amino Acids: Aspartic, Glutamic, Arginine, and Deaminoarginine

Aspartic and glutamic acids contain one carboxyl group in their side chain, which makes them acidic. They have a pK of about 4.4, and at physiological pH they exist as the aspartate and glutamate ions.

Aspartic acid is present in many proteins of biological importance; one of them, glucagon, will be discussed in Chapter 8.

In the urea cycle, the first cyclic metabolic pathway to be discovered (1), one of the nitrogens of the urea is transferred from aspartate.

Aspartic acid is also related to bacterial motion toward some chemicals and away from others, a phenomenon called chemotaxis. As shown by Julius Adler in the 1960s, chemotaxis begins with the binding of the attracted and repelled molecules to receptors called chemoreceptors. These are proteins that bind such attractants as aspartate, while some of the other attractants and repellents are carried by other proteins through the plasma membrane. This process sends signals that result in the forward motion of the bacteria or in the altering of their course. The chemoreceptors found in the plasma membrane are encoded by genes called tsr, tar, trg, and tap. Originally, they were called methyl-accepting chemotaxis proteins. The tar protein binds the aspartate, and it is known as the aspartate receptor. The domain of the receptor that binds the attractant is called the periplasmic domain. Its three-dimensional structure in the aspartate receptor has been determined recently and was found to form an elongated structure (1), with the binding sites for aspartate located more than 60 A from the membrane surface.

The flagellar rotation of bacteria is encoded by the che genes, and the direction is controlled by the CheY protein. The acid pocket of this protein contains three aspartate groups (1).

In order to understand the binding of aspartate to a receptor, it is necessary to study the stability of different conformers of the acid or the ion and the energy cost for the molecule to adopt different conformations.

Sapse et al. (2) used the Gaussian 90 computer program (3) to investigate the potential energy surface of the aspartic acid in its neutral form. The method used was ab initio (Hartree–Fock), with the 6-31G basis set.

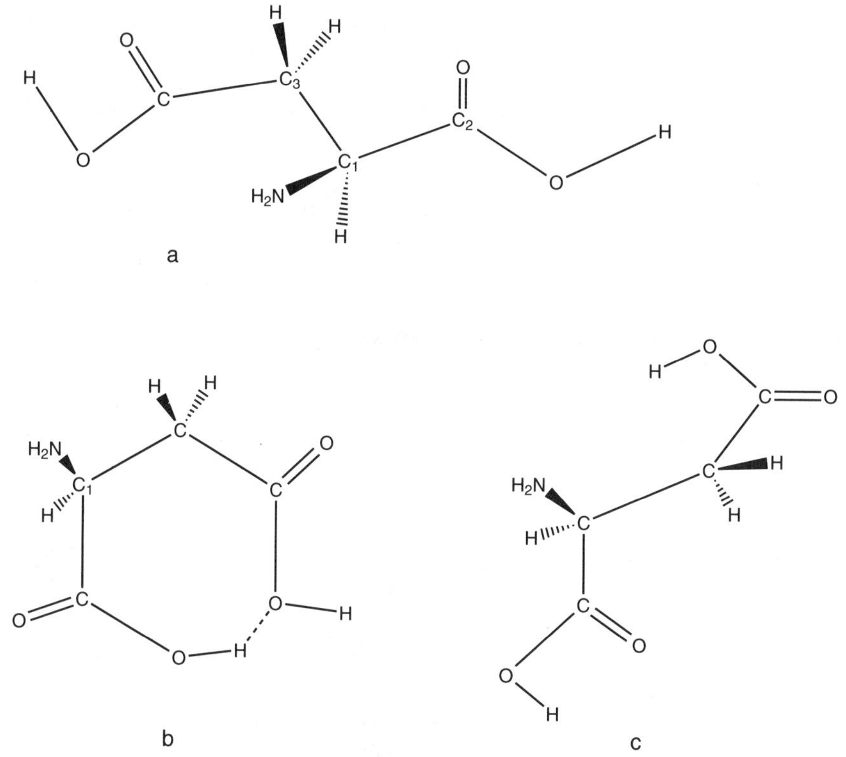

FIGURE 5.1 a. Extended conformer of aspartic acid. b. Cyclic conformer of aspartic acid. c. Partially folded conformer of aspartic acid.

This approach to amino acid structure elucidation has been proven to predict results in agreement with experimentally obtained ones, as in the case of arginine (11). The theoretical values of equilibrium bond lengths and angles, as well as of dihedral angles, were obtained by the Berng optimization method. Different starting geometries were used in order to obtain conformers of aspartic acid, which are either local minima or the global minimum on the potential energy surface, as shown in Figure 5.1. Out of these, the global minimum was found to be the conformer represented in Figure 5.1a. This conformer features an all-extended structure with the C4C3C1C2 dihedral angle of 162.7°. The main carboxyl (set on C1) is rotated only by 7.6° from the C2C1C3 plane, whereas the second carboxylic group makes a 63° angle with the C4C3C1 plane. This structure has an energy of −509.2384 au and is set at zero when the relative energies are calculated.

The structure 5.1b features a hydrogen bond between the second carboxyl hydrogen and the nitrogen of the amino group. It is a local minimum with an energy of -509.2283 au and a relative energy, with respect to 5.1a, of 6.4 kcal/mole. It is interesting to notice that as with the relative energies of the conformers of GABA, the formation of an intramolecular hydrogen bond in the neutral species is not sufficient to stabilize the molecule to the point of making it the global minimum. Structure 5.1c also features an intramolecular hydrogen bond, but this time it is between the main carboxyl's hydrogen and the nitrogen. This conformer features an energy of -509.2337 au and a relative energy with respect to structure 5.1a of 2.9 kcal/mole. It is thus seen to be more stable than structure 5.1b. All the above energies are obtained at Hartree–Fock level, using the 6-31G basis set. The eigenvalues of the Hessian matrix are positive, whence these structure are real minima.

As expected, the O–H bond length involved in hydrogen bonding is longer, with a value of 0.958 A instead of 0.955 A in structure 5.1b and 0.965 A in structure 5.1c.

Glutamic Acid

Glutamic acid differs from aspartic by the presence of an additional CH_2 group. However, as will be discussed in Chapter 8, this small difference can cause large effects in the binding to receptors and activity of proteins in which aspartic residues have been replaced by glutamic residues.

Glutamate is formed in mammalian tissues by the reaction of alanine and ketoglutarate, producing pyruvate and glutamate. This reaction is catalyzed by alanine aminotransferase. From the glutamate ion, ammonium ion is formed by oxidative deamination, catalyzed by the enzyme glutamate dehydrogenase (1).

Glutamate and glutamine play a very important role in the entry of $NH4^+$ ion in amino acids. Glutamate uses an ATP molecule as a source of energy to capture an ammonium ion and form glutamine.

Urea, synthesized in the liver, removes excess $NH4^+$. A blockage of its synthesis leads to elevated quantities of ammonium ion in the blood, which can lead to coma and brain damage. It is not yet known why the ammonium ion is so toxic (1), but a possibility is that elevated levels of glutamine lead to brain damage.

Glutamic acid also plays a role in chemotaxis. Reversible methylation controls adaptation (1), and one of the ways it proceeds is via the methylation of four glutamate side chains present in the chemoreceptors.

As for aspartate, the binding of glutamate to receptors and to active sites of enzymes could be better understood if its structure were determined.

Sapse et al. (2) applied quantum-chemical ab initio calculations to the study of two conformers of neutral glutamic acid. As in the case of aspartic acid, the Gaussian-90 (3) computer program was used to perform

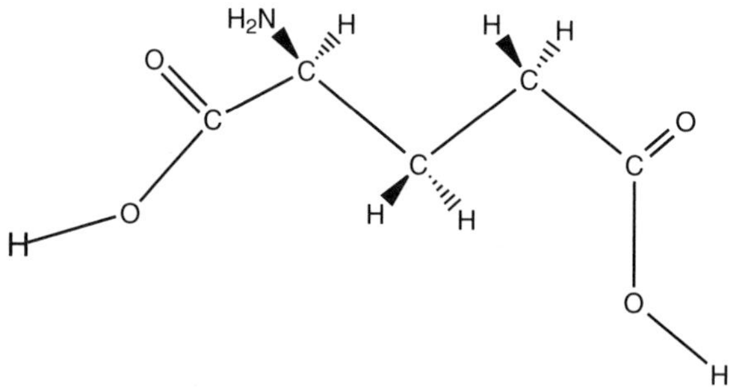

a

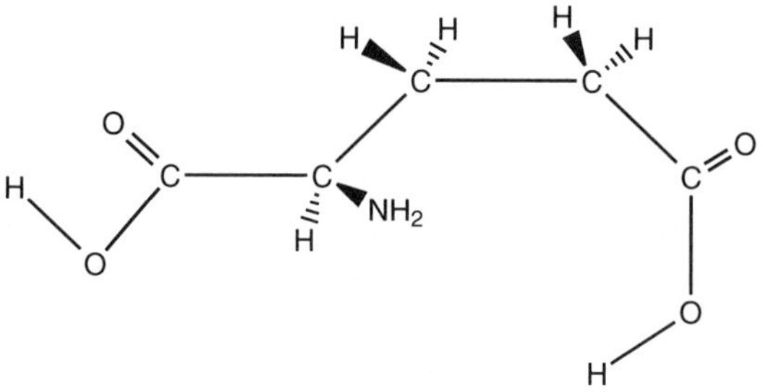

b

FIGURE 5.2 a. Extended conformer of glutamic acid. b. Partially folded conformer of glutamic acid.

Hartree–Fock geometry optimization and energy calculations with the 6-31G basis set. The two conformers investigated, shown in Figure 5.2, are a quasi-extended structure, shown in Figure 5.2a, and a cyclic structure featuring a hydrogen bond between the O–H of the side-chain carboxyl and the nitrogen atom. This structure is called 5.2b. Structure 5.2a is the most stable of the two, with a total energy of −548.2599 au. Structure 5.2b features an energy of −548.2552 au and a relative energy with respect to 5.2a of 2.9 kcal/mole. Structure 5.2a features torsion angles as follows: C4C3C1C2 = 175.1°, O2C2C1C3 = 114.1°, C5C4C3C1 = 71.2°, and O3C5C4C3 = 44.0°. It can thus be seen that this is a partially folded conformer. Structure 5.2b shows a lengthening of the O3H bond length (to 0.971 A), typical of an OH involved in a hydrogen bond.

Investigation of other conformers of glutamic acid showed it to be a flexible molecule.

Arginine and Deaminoarginine

The amino acid arginine is of particular interest because it contains a guanidinium group. The guanidinium group is an example of Y aromaticity, as it was termed by Gund (4), and its interactions with the carboxyl groups of amino acids stabilize the secondary structure of proteins. The guanidinium ion also contributes to protein-DNA recognition via its interaction with the phosphate groups (5).

The structure of the guanidinium ion has been investigated by NMR, MINDO/3, and ab initio methods (6). The rotational barrier around a CN bond was obtained experimentally by NMR methods and found to be 13 kcal/mole. Sapse and Massa (6), using the Gaussian-70 computer program with the 6-31G basis set, found a value of 14.79 kcal/mole for a single rotation, 43.34 kcal/mole for a double rotation, and 111.63 kcal/mole for a triple rotation. However, that work does not use a large enough basis set and does not take into consideration correlation energy effects. Therefore, it is of interest to recalculate the rotational barriers of the guanidinium ion at a much higher calculational level. Therefore, calculations were performed (7) for the optimization of the guanidinium ion, using the Moller–Plesset method (MP2) with the 6-31G** basis set, which sets d polarization functions on the nonhydrogen atoms and p polarization functions on the hydrogens. Surprisingly, the rotation energy around a single bond was 14.53 kcal/mole, very close to the result of 14.73 kcal/mole (6) obtained with the 6-31G basis set, without polarization functions and correlation energy calculations. It might be concluded that a double-zeta set is adequate for the study of guanidinium.

Herzig et al. (8) presented a study of substituted guanidinium ions in which one of the hydrogens was substituted by either a fluorine ion or an amino group or a methyl.

The choice of these substituents was made on the basis of their electron-donor acceptor properties, which for these three possibilities cover a large range. Experimental data on the rotational barriers of substituted guanidinium have been obtained by NMR (9). Those results indicated changes of few kcal/mole for the rotational barriers, changes rationalized in terms of substituent interaction with the Y aromatic charge distribution of the guanidinium ion. Another factor that was found to influence the rotational barrier changes was intramolecular hydrogen bonding. Substituents' effects on the charge distribution of the ion indicate how they will affect the hydrogen bond formation.

Herzig et al. (8) use the 6-31G basis set to optimize (subject to some constraints) the geometry of the system and to calculate the energies in ground state and after rotation. They use the Hartree–Fock method, without correlation energy calculations. For all three substituents involved, the framework of the planar guanidinium is preserved, with CN bond lengths of 1.33 Å and planarity and HNH angles of 120.0°. The FNC angle features a value of 115.0°, which brings it in proximity to the hydrogen on one of the other nitrogens, leading to a formation of a hydrogen bond with a F. .H distance of 2.12 A. The F–N bond length is 1.36 A. The amino substituent features a pyramidal geometry with the lone pair of nitrogens directed toward a hydrogen on one of the other nitrogens. Again, there is the formation of a hydrogen bond, with a length of 2.34 A. The N–N distance is 1.42 A. For the methyl substituent, the C–N bond is 1.48 A.

The rotational barriers around the three CN bonds are not the same since the symmetry has been removed. Thus, the rotation around the CN bond in which the nitrogen holds the fluorine is 19.70 kcal/mole, the rotation around the CN bond cis to the fluorine is 19.31 kcal/mole, and the third is 11.53 kcal/mole. Indeed, the first two rotations in addition to breaking the pi bond break the hydrogen bond between the fluorine and the hydrogen. The electronegative fluorine localizes more of the pi charge on the CN to which it is attached, leading to a rise in the barrier.

In the amino substituent case, the same effect is observed, with the rotation around the CN attached to the amino group 23.40 kcal/mole, the rotation around the adjacent CN 22.79 kcal/mole, and the one trans to the amino group 12.97 kcal/mole. Again, the breaking of the hydrogen bond is reflected in the size of the barriers. In the case of the methyl substituent, the changes are much smaller, since the methyl group resembles hydrogen the most in electronegativity.

A study by Sapse et al. (10) has applied the same methods of calculation to the rotational barriers around the CN bonds in guanidine and substituted guanidine. Of course, in guanidine the CN bonds are not equivalent as they are in the guanidinium ion. Figure 5.1 shows guanidine and the substituted guanidines. Through Hartree–Fock calculations, using the 6-31G basis set, it was found that the bond length for the CN bond where N is the imino nitrogen is 1.29 A. As the NH rotates to a 90° angle around this bond, the length becomes 1.32 A. When there is rotation around the other two CN bonds, their bond lengths increase from 1.37 A to approximately 1.4 A. The value of 1.37 A is intermediate between a CN single bond and a CN double bond. When rotation occurs, the pi character of the bond decreases and the bond becomes closer to a single bond; therefore, its length increases.

Examining the values of different rotation barriers one can conclude that, as expected, the pi electron delocalization in guanidine still exists even though it is much less than in guanidinium ion.

Arginine and deaminoarginine have also been studied with ab initio methods in a study performed by Sapse and Strauss (11).

Hartree–Fock calculations, using the STO-3G minimal set as implemented by the Gaussian-82 computer program, were used to calculate the optimum geometry of the global minimum conformation as well as of some local minima conformations of arginine and deaminoarginine.

For each of these three molecules, three conformers were investigated:

- An all-extended structure.
- A cyclic structure featuring a hydrogen bond between the hydrogen of the carboxyl group and the nitrogen of the guanine entity.
- A partially folded structure.

These conformers are shown in Figures 5.3 (arginine) and 5.4 (deaminoarginine).

In addition, a zwitterion that features a cyclic conformation with H1 set on the guanidine's nitrogen and hydrogen bonded to the carboxylate group was also investigated.

The cyclic structures were given initial torsion angles that ensured the formation of the hydrogen bonds. The partially folded structures feature extended conformations for all the carbons except the carboxyl's carbon, which was set in a gauche conformation.

At STO-3G level, it was possible to optimize all the parameters of the molecule. However, with the computation facilities available when this study was undertaken, the 3-21G complete optimization was cumbersome. In consequence, fragments (such as the guanidine molecule, the guanidinium ion, acetic acid, and the acetate ion) were optimized at 3-21G level, and the STO-3G optimized parameters were used for interfragment bond lengths and angles. The rationale for this procedure was that the STO-3G basis set is not adequate for the description of zwitterions (12). Accordingly, the 3-21G calculated energy obtained from the guanidinium and acetate ion entities was compared with the 3-21G energy obtained from guanidine and acetic acid entities. In order to examine the possible existence of a barrier to the proton transfer from the carboxyl to the guanidinium group, the NH1 distance was set at 1.2 A, instead of the 1.05 A distance present in the zwitterion.

It was found (11) that the most stable conformations both for arginine and deaminoarginine are the partially folded ones, similar to the results obtained for GABA (see Chapter 3). They were closely followed by the extended structures. This result is in agreement with experimental crystallographic data that show the carboxyl group either trans or gauche to the rest of the molecule (13). The least stable conformations are the cyclic ones, again similar to GABA.

The 3-21G calculations show the zwitterion to be a local minimum, higher in energy than the neutral cyclic structure by 16 kcal/mole. A barrier of 29 kcal/mole was found, corresponding to the energy of the structure featuring NH1 = 1.2 A.

The binding interactions of aspartic and glutamic acid carboxyl groups in the hydrophobic interior active sites of proteins occur with basic protein groups or with other ligands. The carboxylates can thus interact with the imidazole group of histidine as in chymotrypsin, the amino group of lysine, or the guanidinium group of an arginine residue. The interaction might be between the carboxyl group of a ligand and the basic group of the enzyme

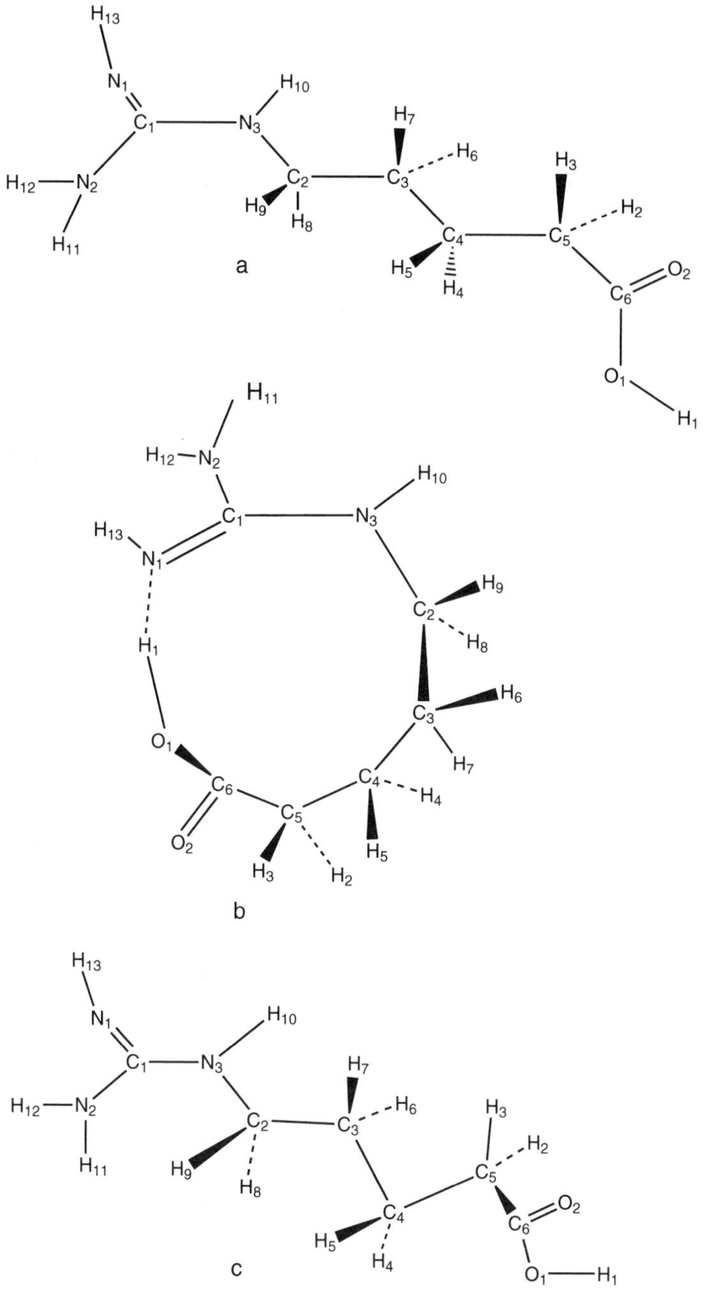

FIGURE 5.3 a. Extended conformer of deaminoarginine. b. Cyclic conformer of deaminoarginine. c. Partially folded conformer of deaminoarginine.

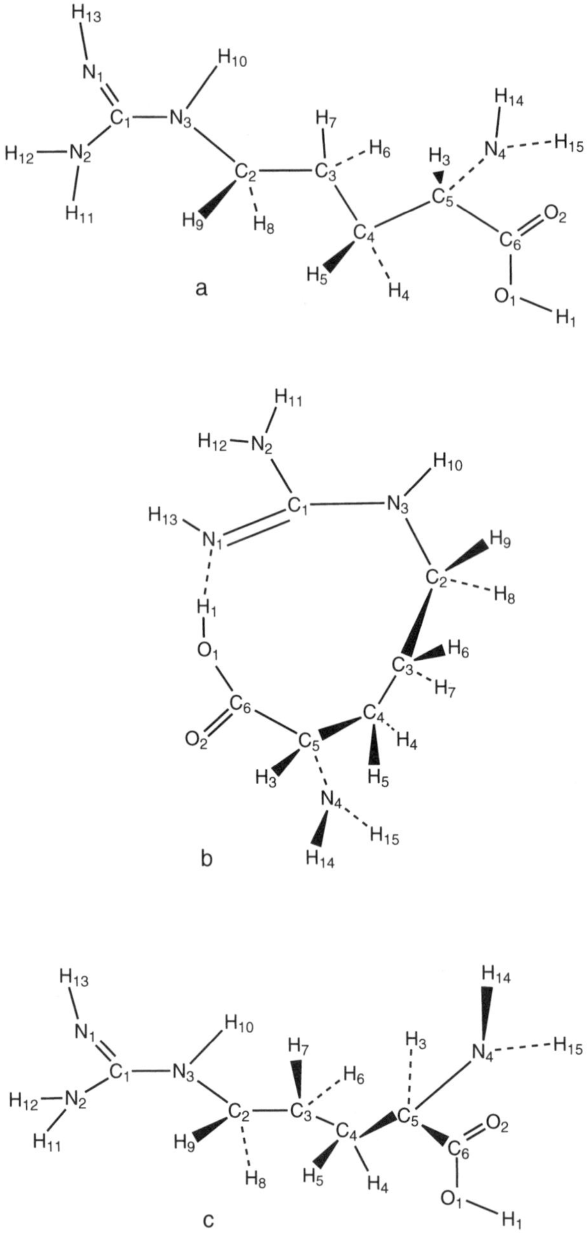

FIGURE 5.4 a. Extended conformer of arginine. b. Cyclic conformer of arginine. c. Partially folded conformer of arginine.

or vice versa. For example, Andreevna et al. (14) observed an interaction in pepsin between the N-terminal amino group and the carboxyl of Glu4, which brings the N-terminus into the core. They also observed a conserved aspartic 11 carboxyl-arginine 308 guanidinium interaction. Jansonius et al. (15) have proposed that in the reaction of aspartate aminotransferase with the substrate, the two carboxyl groups of the substrate are interacting with the guanidinium groups of arginine 386 and arginine 292 of the enzyme. These interactions orient the substrate for catalysis, whereas the amino group of the lysine 258 acts as proton acceptor–donor in the formation of a ketimine intermediate.

One of the interesting properties of arginine is that it might form a tighter ion pair with a carboxylate group than does lysine (16). These ion pairs do not break in beta or gamma turns occurring in natural enzymes.

Sapse and Russell (17) have applied theoretical methods to the study of the binding of methylamine and guanidine to carboxylate. The main purpose of that work was to compare the energy of the neutral complex to that of the zwitterion, focusing on the proton transfer between the guanidinium and formate ions.

The systems investigated were the following:

• Guanidine and formic acid;
• Methylamine and acetic acid;
• The guanidinium ion and the formate ion.

The 6-31G basis set was used for Hartree–Fock calculations of the energies of the guanidinium–formate ion complexes. The subspecies, namely the guanidinium ion, guanidine, formic acid, and the formate ion, were subjected to complete geometry optimization. The complexes were also geometry optimized. For the zwitterion, an energy minimum was found, with all the eigenvalues of the second derivative matrix positive, so that it is a real minimum. When the proton is moved away from the guanidinium entity toward the formate ion, at such distances as N–H = 1.2 A and 1.4 A, no stationary point is found.

In the case of the methylamine–acetic acid complex, calculations were performed with the 6-31G, 6-31G*, and 3-21G basis sets. The systems investigated were the following:

• The protonated methylamine–acetate ion complex;
• The methylamine–acetic acid complex;
• The methylamine–acetate complex with the proton set at a distance of 1.5 A from the amine's nitrogen.

All the above complexes are doubly hydrogen bonded. In addition, a single hydrogen-bonded complex was investigated, with a geometry that features a linearity of the N–H–O bond.

The 3-21G calculations show a stationary point with all the eigenvalues of the second-derivative matrix positive for the methyl amine–acetic acid complex. The zwitterion is not a stationary point. At both the 6-31G and 6-31G* levels, the lowest energy is found for the complex of methylamine and acetic acid.

Barriers were not found for the proton transfer in either the guanidinium or in the methylamine complex. When the proton is transferred from the most stable structure to the least stable one, there is a gradual increase of the energy.

References

1. Stryer, L. *Biochemistry*, 4th ed. W.H. Freeman, and Company, New York, 1995.
2. Sapse, A.M., Mezei, M., Jain, D.C., and Unson, C. *THEOCHEM 306*, 225, 1994.
3. Frisch, M.J., et al. *Gaussian 90*, Gaussian Inc., Pittsburgh, PA, 1990.
4. Gund, P. *J. Chem. Ed. 49*, 100, 1972.
5. Eggleston, D.S., and Hodgson, D.J. *J. Peptide Protein Res. 25*, 242, 1985.
6. Sapse, A.M., and Massa, L.J. *J. Org. Chem. 45*, 719, 1980.
7. Sapse, A.M. Unpublished results.
8. Herzig, L., Massa, L.J., Santoro, A.V., and Sapse, A.M. *J. Org. Chem. 46*, 2330, 1981.
9. Santoro, A.V., and Mickevicius, G. *J. Org. Chem. 44*, 117, 1979.
10. Sapse, A.M., Snyder, G., and Santoro, A.V. *J. Phys. Chem. 85*, 662, 1981.
11. Sapse, A.M., and Strauss, R. *Int. J. Quant. Chem. XXXVII*, 197, 1990.
12. Fugler, L., Russell, C.S., and Sapse, A.M. *J. Phys. Chem. 91*, 37, 1987.
13. Coll, M., Subirana, J.A., Solans, X., Font-Altaba, M., and Mayer, R. *J. Pept. Protein Res. 29*, 708, 1987.
14. Andreevna, N.S., Zdanov, A.A., and Fedorov, A.A. *FEBS Lett. 125*, 234, 1981.
15. Kirsch, J.F., Eichele, G., Ford, G.C., Vincent, M.G., Jansonius, J.N., Gehring, H., and Christen, P. *J. Mol. Biol. 174*, 497, 1984.
16. Wigley, D.B., Lyall, A., Hart, K.W., and Holbrook, J.J. *Biochem. Biophys. Res. Comm.* 149, 927, 1987.
17. Sapse, A.M., and Russell, C.S. *J. Mol. Structure (THEOCHEM) 137*, 43, 1986.

6
Proline

Proline features the amino group as part of a five-membered ring, differing thus from other amino acids in that its side chain is bonded to the backbone nitrogen atom and to the alpha carbon atom. This cyclic structure influences greatly the protein architecture.

It is unique among amino acids because it features only one hydrogen bound to the amino nitrogen. The five-membered ring of proline prevents rotations about the alpha carbon atom and the nitrogen bond, restricting thus the phi angle to a value of about 65°. This reduces greatly the number of conformations available to proline. Moreover, the steric hindrance introduced by the five-membered ring restricts the conformation available to the residue at the amino terminal of proline. Specifically, in an X-Pro sequence, there is little probability that the X will be an alpha helix. Proline itself disfavors an alpha helix because it does not have an amide H atom that could form hydrogen bonds.

Another specific property of X-Pro sequences is that about 6% of them are cis, as opposed to other peptide bonds, which are usually trans. Because of this property, as will be discussed in Chapter 10, proline is prone to the formation of tight turns. Prolyl isomerization is the rate-limiting step when proteins are folding in vitro (1). Indeed, this step is slow because the carbonyl carbon and the amide nitrogen form a bond featuring a partial double character. The enzyme peptidyl prolyl isomerase lowers the activation barrier for cis–trans isomerization by twisting the peptide bond and allowing it to feature more of a single-bond character. The isomerization is accelerated thus 300-fold.

Increasing the content of proline and hydroxyproline in some proteins, such as collagen, increases their melting temperature (1). Hydroxyproline is formed by the hydroxylation of prolines in the amino side of glycine in nascent collagen structures by using one oxygen from O_2. The other oxygen from O_2 is taken from alpha-ketoglutarate, which is transformed into succinate. The enzyme that catalyzes this reaction is prolyl hydroxylase, assisted by an Fe_2^+ ion.

Proline is degraded in the organism by entering the citric acid cycle at alpha-ketoglutarate. It is converted into glutamate gamma-semialdehyde, which is subsequently oxidized to glutamate. An intermediate in this reaction is pyrroline 5-carboxylate.

The structure of proline has been studied by experimental and theoretical methods. Among the experimental methods are NMR, IR, and CD spectroscopy (2–4). Among the theoretical methods used in the study of proline are molecular mechanics (5), PCILO (6), and ab initio methods.

Cabrol et al. (6) applied PCILO calculations to Ac-Pro-Pro-GlyMe2-Ac and found that a C7 ring is the most stable structure for the endo proline ring. The ORD studies of Scheelman and Neilson (7) also indicate the C7 ring to be the most stable structure for N-Acetyl-N'-methylproline amide with an endo proline ring.

Brunne et al. (8) applied molecular-dynamics simulations to the proline-containing cyclic peptide antamanide in order to compare the conformational equilibrium and dynamics of four proline residues with NMR results. The purpose of the work is to test whether the general force field of GROMOS (9) is able to lead to results in agreement with the experimentally obtained conformational equilibrium distribution and to provide a correct description of the conformational dynamics. In order to accomplish this task, Brunne et al. (8) have chosen the phenomenon of proline ring flips within the cyclic decapeptide antamanide, which contains four proline residues and for which experimental data could be found.

Antamanide's structure as well as its backbone and side-chain rotations (10–12) have been studied in solution, using various methods. A detailed analysis of the NMR data referring to the proline flips was presented by Madi et al. (13). They conclude that Pro2 and Pro7 oscillate rapidly between two energetically alike conformations, while Pro3 and Pro8 feature one conformation only.

Brunne et al. (8) use the stochastic dynamics (SD) method (14), in which they can include the influence of the nonpolar solvent in the frictional and random forces of the Langevin equation of motion, without treating the solvent molecules explicitly. Indeed, it has been shown (15) that stochastic dynamics describes well the solute–solvent interaction in the case of nonpolar solvents.

Brunne et al. (8) report that the observed proline flips cause all the torsional angles between residues to change simultaneously. They group Pro2 and Pro7 in one behavioral group that exhibits transitions between two conformations, remaining in each for several picoseconds. Pro3 and Pro8 have to undergo transitions between a stable and a high-energy conformation. The latter features a lifetime of about 1 picosecond.

The two conformations of each proline are characterized mainly by the chi2 angle, which describes the puckering of the proline ring. As such, positive chi2 defines one type of conformation, while negative chi2 define the other. Madi et al. (13) using NMR methods, have determined the values for the correlation times for the proline puckering of Pro2 and Pro7 to be of order 30 and 36 ps, respectively. These numbers, as shown by the authors, have a large uncertainty. Brunne et al. (8) obtain smaller residence times from molecular-mechanical simulations.

As pointed out by Hurley et al. (16), accurate conformational energy maps are of practical importance not only for theoreticians but also for experimentalists involved in the engineering of thermostable proteins. Any new set of empirical potential energy functions is supposed to reproduce the known conformational energy map (17–18). Proline, due to the special restrictions imposed by the presence of the ring, plays a special role in the conformations of proteins. This role is not related only to the proline residue but also to the conformation of the residue preceding the proline. Calculations performed by Summers and Karplus (19) and Schimmel and Flory (20), based on ideal covalent bond lengths and angles, have shown that the calculated energy of the alpha conformation of the residue that proceeds a proline in a peptide features an energy higher by more than 7 kcal/mole than the beta conformation.

As pointed out by Hurley et al. (16), this difference is higher than the free energy of folding of many proteins, so the alpha conformation of a residue before a proline might never be present. Indeed, even though proline occurs with approximately 26% frequency in alpha and 3_{10} helices, it is usually in the first helical position. However, it happens in about 9% of the cases that non-Pro and non-Gly residues preceding proline are in a helical conformation (21). To replace a non-Pro residue by a Pro in an alpha helix involves an energy cost combining the effects of hydrogen-bond breaking and of conformational changes that destabilize the helix. Alber et al. (22) have introduced a number of such site-directed mutations in T4 lysozyme. O'Neil and DeGrado (23) replaced an alanine residue in an oligopeptide by a proline residue and found that the energy increased by 3.4 kcal/mole. This energy is smaller than the 7 kcal/mole that is the cost in energy relative to the preceding residue of a proline. In order to further investigate the reasons for possible discrepancies between calculational results and experimental results, Hurley et al. (16) performed energy calculations on an ala-ala-pro tripeptide.

The Amber all-atom force field is used in that work to carry out the calculations. The choice of the ala-ala-pro peptide as a model was made in order to study the effects of neighboring residues on conformational energies. The solvent effect is taken into consideration by calculating the stereochemical and Lenard–Jones energy terms with a dielectric constant epsilon = 80 (for water) (16). The N and C termini of the oligopeptide were neutral. Energies were calculated as a function of the phi and psi angles of Ala2, with 10 increments in the angles. The bond lengths and bond angles were kept frozen at ideal values (24). While Ala1 was kept in a helical conformation, with psi = $-47°$, calculations were performed for Pro3 in both extended and helical conformations.

Another set of calculations allows bond lengths and angles to be optimized (16) using the method called "flexible geometry" (24).

For the rigid geometries maps, the results obtained by Hurley et al. (16) resemble those obtained previously (19,20). The difference in energy between the alpha and beta conformations is 10.2 kcal/mole. The conformation of Pro3 is found to have a negligible influence on the conformation of the preceding residue. The flexible geometry maps show much shallower minima and much smaller differences between the alpha and beta conformations. The difference in energy between the most stable conformation in the helical region and the overall minimum energy conformation is only 1.1 kcal/mole. The flexible geometry results are in good agreement with the distribution of conformations in known structures. The 1.1 kcal/mole difference between alpha and beta conformations is possible, given the approximate 9% frequency of the alpha conformations. The results are also plausible when experimental stability data on mutations that add or remove prolines in helical peptides are examined (16).

Proline replacement at solvent-exposed sites in alpha helices changes the stability by less than 4 kcal/mole (19). This is less than the 7 kcal/mole value obtained from rigid geometry maps. Since the conformational energy alone is only a fraction of the 4 kcal/mole, the 1.1 kcal/mole value is reasonable.

Yun et al. (26) obtained a value of 3.4 kcal/mole for the increase in the free energy of folding for a proline in an alpha helix. This result was obtained by using the free-energy perturbation method (FEP) in an explicit solvent model. This method includes both hydrogen bonding and conformational energy contributions (16). Accordingly, a map of the conformational energy alone is reasonable if the energy difference between the extended and helical conformations is much less than 3.4 kcal/mole. This is additional proof that the flexible geometry map results are more credible than the rigid geometry map results.

Although the energies are consistent with experimental results in the flexible geometry model, there are some discrepancies in the values of some parameters as calculated and as found in the Pro86 in the T4 lysozyme (27). These discrepancies could be related to an imbalance between the bending and stretching force on one side and the steep nature of the $1/r$ repulsive potential used in most molecular-mechanical force fields (16).

Hurley et al. (16) conclude that the results of this type of calculation are much improved by allowing all the parameters of the molecule to relax.

Ab initio calculations on proline and systems including proline were performed by a number of researchers. All these calculations include optimization of all the parameters of the system.

Sapse et al. (28) applied the ab initio method to the proline amino acid itself, to N-formylprolyne amide, and to N-acetylproline amide. They used the Gaussian-80 computer program with the 6-31G and STO-3G basis sets to perform Hartree–Fock calculations. The geometry was optimized with the Berny optimization method, except for the angles $NC^\alpha C^\beta = NC^\delta C^\gamma$ and $C^\alpha NC^\delta$, which were subjected to point-by-point optimization.

In proline, the carboxyl group was set cis to the N–H bond. The STO-3G basis set was used to optimize the angles mentioned above with a grid search in increments of 0.5. The values thus obtained for the three angles were used for the 6-31G optimization of the molecule. Both bond lengths and angles compare well with experimental values (28). However, the ring puckering that was investigated by calculating the optimum value of the C^β $C^\alpha NC^\delta$ and $C^\gamma C^\delta NC^\alpha$ dihedral angles differs from the experimental values both for the STO-3G and 6-31G values. This is due to the fact that for a correct description of the ring puckering, polarization functions are necessary. However, the computational task required for calculations involving polarization functions for a relatively large system at that time was too big. The differences between the 6-31G and the STO-3G-predicted geometries were small. Accordingly, the N-formylproline amide was investigated with the STO-3 basis set. The formyl group was supposed to be coplanar with the proline ring in order to ensure maximum pi overlap. The N thus has sp^2 character. The $NC^\alpha = NC^\delta$ constraint used for proline was removed. The structure examined featured a hydrogen bond between the oxygen of the formil group and one of the hydrogens of the amide.

During the geometry optimization of N-acetylproline amide, the CNC angles were kept at the values obtained for N-formilproline amide. The trans peptide conformations of the molecule were investigated, and it was found that the most stable structure features a hydrogen bond between the oxygen of the acetyl group and one of the hydrogens of the amide group. The structure second in stability does not feature this hydrogen bond and has the nitrogen of the amide group above the plane of the ring.

A structure that features the oxygens of the acetyl group and of the amide in proximity was found to be quite high in energy.

The ab initio–obtained parameters of N-acetylproline in its most stable conformation are in good agreement with the experimental values found by Karle (29) for crystalline cyclo Gly-D-Ala-Pro-Gly-Pro, specifically those of the center proline. The structure next in stability is close to the crystal structure of N-acetylproline methylamide investigated by Matsuzaki and Iitaka (30), which contains intermolecular rather than intramolecular hydrogen bonds.

Sapse et al. (28) found the peptide dihedral angle in N-acetylproline amide to be twisted slightly out of plane, in agreement with the experimental results of Karle (29). Since the structure slightly higher in energy does not feature this twisting, it is suggested (28) that the twisting might be due to the formation of the hydrogen bond. The phi and psi angles are in agreement with the results of Cabrol et al. (6).

This work does not take into account solvent effect. As such, the results are expected to agree best with the experimental results obtained in a nonpolar solvent, such as carbon tetrachloride (31). Indeed, in this solution the hydrogen bond is present, while in crystals (30) it is absent. The calculations cannot take into account intermolecular hydrogen bonding, so it is reasonable that the nonhydrogen-bonded conformation is higher in energy than the hydrogen-bonded one. The difference in energy between these two conformations is found to be 3.7 kcal/mole at STO-3G calculational level (28).

Sapse et al. (28) conclude that N-acetylproline amide forms a 1:3 hydrogen bond, with an energy of about 3.7 kcal/mole and with a distance between the hydrogen and the oxygen involved in the bond of 1.8 A. The cis conformation is higher in energy by 4.1 kcal/mole, which makes it higher than the trans conformation without a hydrogen bond by only 0.4 kcal/mole. This value agrees with the molecular-mechanics results of DeTar and Luthra (5) for the cis and trans transformations of N-acetylproline methyl esther. It also agrees with the results of Pullman et al. (32), who find a 0.5 kcal/mole stability of the trans nonhydrogen-bonded isomer over the cis isomer. It is thus clear that ab initio calculations using an STO-3G basis set can predict the existence of a 1:3 hydrogen bond, even though the puckering of the ring is not accurately predicted.

Ramek et al. undertook studies of N-acetylproline amide using ab initio methods with a number of basis sets. They investigated the potential energy surface (PES) in terms of local minima and transition states (33).

Ramek et al. (33) used the GAMESS program to perform restricted Hartree–Fock calculations (RHF) with the 4-31G and the 6-31G* basis sets. The RHF calculations resulted in the finding of seven local minima, which they label cYz. In this notation, $x = c$ or t represents the cis or trans orientation of the acetyl group, referring to the C^m–C–N–C^α torsion angle. $Y = A$, B, C, F describes the orientation of the amide group, according to the labeling method introduced by Zimmerman et al. (34). The notation $z = u$ or d describes the exo or endo puckering of the proline ring.

$Y = A$ describes conformations with $-110° \leq \phi \leq -40°$ $-90° \leq \psi \leq -10°$;
$Y = B$ describes conformations with $-180° \leq \phi \leq -110°$ $-40° \leq \psi \leq -20°$;
$Y = C$ describes conformations with $-110° \leq \phi \leq -40°$ $50° \leq \psi \leq -130°$;
$Y = F$ describes conformations with $-110° \leq \phi \leq -40°$ $\psi \geq 130°$ $\psi \leq -140°$.

Three of the local minima feature an endo puckering of the proline ring, while the other four feature an exo conformation. The endo conformations are lower in energy than their exo analogues.

The difference in energy between the endo and their exo analogues is 4–8 kcal/mole. The tAu conformer has a very small potential barrier for the CONH2 rotation. However, the tAd conformer has no barrier.

Ramek et al. (33) found the global minimum to correspond to the tCd conformer. This conformer and its pucker-up analogue are stabilized by an intramolecular N–H ... O=C hydrogen bond, with a H ... O distance of slightly more than 2 A.

The *cBd*, *cAu*, and *tAu* conformers feature another interaction, of the form N–H . . . N. However, since there is no critical point in the electron density along the connection line, this interaction is not a hydrogen bond (33).

Ramek et al. (33) also investigated the reaction paths between different conformers. They found that the local minima are interconnected in the following way:

- The endo and exo analogues are connected.
- For each reaction of conversion between puckering-down conformers there are two paths corresponding to clockwise and anticlockwise internal rotations (except for *tCd* ⇌ *tCd*, for which these two directions are the beginning and the end of the same path).
- The pucker-up conformations feature only two reactions with a distinct saddle point: *tCu* ⇌ *tAu* ⇌ and *tAu* ⇌ *cAu*.
- Each of the seven local minima can undergo internal rotations of the –CH₃ and –NH₂ groups in which the starting structure is retrieved.

When the results obtained via 4-31G and 6-31G* basis sets are compared, two essential differences are noticed. The 6-31G* basis set predicts a C=O bond length of approximately 0.02 A less than the one predicted by the 4-31G basis set. Also, the N–H . . . O=C hydrogen bonds are longer by 0.01 A, due to the increased value of the C=O bond length. The second difference is the fact that via 4-31G calculations the NH₂ group is almost coplanar with the adjacent C=O group; for all conformers, the 6-31G* basis set predicts nonplanarity for these two groups in the case of *tCd*, *tCu*, and *tAu*. As a result there is a difference of about 15° between the O=C–C–N and N–C–C–N torsion angles, as predicted by the two basis sets.

The relative energies and the potential barriers also show some differences between the values obtained by the two sets. For example, the conformer *tAu* has its relative energy reduced by about 40% when the 4-31G basis set is changed to 6-31G*. All other conformers have their relative energies reduced, with values between 15% and 20% (33).

Kelterer et al. (35) show that the *tAu* conformation of N-formylproline amide is a local minimum with many basis sets, including 4-31G and 6-31G**, but not with the 6-31G* basis set. Ramek et al. (33) find that the vibrational zero-point energy of this conformer of N-acetylproline amide is below the potential barrier of the *tAu–tCu* reaction, when calculated with the 4-31G basis set, but is above it with the 6-31G* basis set. In general, they find many potential barriers to be decreased from the 4-31G basis set to the 6-31G* basis set.

When Ramek et al. (33) investigated the puckering of the ring with the STO-3G basis set, they found the ring to be puckered, with energies lower than that found by Sapse et al. (28). They attribute this difference to constraints imposed in (28), but it is probably due to the improvement in computer program effectiveness.

Ramek et al. (33) studied these systems also by semiempirical methods.

Using the AM1 method, they found the PES of N-acetylproline amide to contain three local minima. The proline ring, as described by the AM1 method, is almost planar, so endo and exo isomers do not coexist. The conformers are thus called tC, cF, and cA. The first two are identical with the transition states in the $tCd \rightleftharpoons tCu$ and $cFd \rightleftharpoons cFu$ transitions (33). The last conformer is different from the others, since it features a six-membered ring cyclic fragment, with a N–H . . . C interaction. In agreement with the ab initio results, the AM1 results also show the global minimum to correspond to the conformer exhibiting a N–H . . . O=C hydrogen bond.

Another semi-empirical method that Ramek et al. (33) used is the PM3 method. This method predicts the existence of 17 local minima for N-acetylproline amide. These calculations predict that the proline ring will be puckered, thus giving rise to endo and exo conformers. The $-CONH_2$ and the $-CH_3$ groups can exist in two different conformations each (33). To label each conformer, in addition to the RHF nomenclature, two other indices have to be added.

The global minimum is also found to be a conformer with the N–H . . . O=C hydrogen bond. However, as predicted by the PM3 method, this bond is weak, with a H . . . O distance of 2.6 A. In agreement with the AM1 results, the PM3 results also predict the existence of a conformer with a six-membered ring.

Ramek et al. (33) conclude that N-acetylproline amide features three endo conformers, as predicted by RHF calculations. These are tCd, cBd, and cFd. They have exo analogues with barrier potentials of about 10 kcal/mole for the exo–endo conversion. The tCd conformers, which feature a seven-membered ring, is of high kinetic stability (33), while the tAu conformer, which as part of a peptide would feature a ten-membered ring, is of low kinetic stability.

The comparison with experiment, as pointed out by Ramek et al. (33), is difficult, since most experimental data are obtained in the crystalline state (36). The crystalline states of N-acetylproline amide and N-acety-N'-methylproline amide (37–38) exhibit tAu conformations. In larger polypeptides seven-membered rings have also been observed in solid state (39). In polyprolines, where there are no hydrogens from the amino group that could form hydrogen bonds, the cFd and cFu structures are seen (36).

The experimental structures show bond lengths and angles in good agreement with the RHF optimized geometries (33). While the endo puckering is also in good agreement, the experimental exo puckering is larger as found by experiments than its calculational value.

The AM1 method does not reproduce well the puckering of the ring but predicts correctly the stability of the seven-membered cycle conformer. It also replaces the N–H . . . N' five-membered ring conformation by a six-membered ring (33). Ramek et al. (33) found that the PM3 semiexperimental method predicts better results on systems with intermolecular hydrogen bonds.

In a subsequent study, Ramek et al. (40) investigate L-proline at a higher level, using the 6-311++G** basis set, which, besides adding polarization functions on the nonhydrogen and hydrogen atoms, also uses diffuse functions on both type of atoms. They discuss these results with respect to the parametrization of the MM3 force field for molecular-mechanical calculations and compare them to those obtained for glycine, alpha-alanine, and previous results for N-acetylproline.

The ab initio calculations performed by Ramek et al. (40) make use of the GAMESS program, computing the Hessian matrix by double numerical differentiation of analytical first derivatives. They use a stochastic search routine in order to find the local minima within the MM3 (94) force field program. From the located minima, the potential energy curves were obtained using diagonal Newton–Raphson minimization (40).

Ramek et al. (40) found 10 conformers on the potential energy surface of neutral proline, using the Hartree–Fock calculations. The optimization was carried out by a preliminary investigation with the 4-31G basis set, the results of which were used for a 6-31G* search, followed by a final 6-311++G** optimization. This method reveals the dependence of the order of stability on the basis set used. Indeed, the first two conformers in order of stability exchange places as the global minimum as the basis set is changed. The torsion angle H–N–C–C decreases as the size of the basis set increases. The trends show the inclusion of diffuse functions to be of minor importance for the energy but important for structural details. The two most stable conformers do not feature an O–H . . . N hydrogen bond, which is present in the next two conformers (40).

When reaction paths are investigated, Ramek et al. (40) found that except in those that involve an internal rotation of the –OH group, the internal rotation of the –COOH group is coupled with either an inversion of the amino group or with a ring pucker change.

When the MM3 force field is applied, Ramek et al. (40) used the following definition of the torsional energy (the largest energy component for proline):

$$E_T = \frac{V_1}{2}(1+\cos\omega) + \frac{V_2}{2}(1-\cos 2\omega) + \frac{V_3}{2}(1+\cos 3\omega)$$

with ω the angle for four bonded atoms and V_1, V_2, V_3 torsional constants.

Eight conformers are found to be minima with the syn-periplanar C=O and 0–H groups, with two of them similar to the ab initio results. However, the parameters used in the calculation do not define the global minimum correctly. Therefore, Ramek et al. (40) point out the necessity of using an ab initio–determined parameter for the N–C–C–O angle. Since in proline this angle is coupled with the amino group inversion and the ring pucker motions, the value obtained for alanine is used (40). This method ensures agreement between the MM3 global minimum and the ab initio global minimum.

As pointed out in (4o), a result of interest is the strength of the hydrogen bond N . . . H–O, present in some of the stable conformers. In proline, this bond is part of a five-membered ring that is almost planar. Similar bonds in glycine and alanine exhibit bond orders of 0.019 and 0.023, respectively. It can thus be seen that it is stronger for alanine, even though glycine has the lower sterical strain (40), since in the case of proline it was found (40) that the conformer with an eclipsed conformation has a stronger hydrogen bond than the one with a staggered conformation. This result explains the glycine-alanine trend by steric effects (40).

In what concerns the order of stability for the free acid and the amide, Ramek et al. (40) found that in amide the conformers featuring the N–H . . . O=C hydrogen bond are more stable, since the formation of a seven-membered ring stabilizes the system.

In conclusion, Ramek et al. state (40) that various set dependencies are observed in the energetics of the conformers, as well as in the value of some structural parameters. They note the importance of introducing diffuse functions. The MM3 calculations cannot be performed correctly with straightforward parameter guesses. Parameters have to be chosen based on ab initio results. Even then, not all the properties of the system can be predicted.

References

1. Stryer, L. *Biochemistry*, 4th ed. W.H. Freeman and Company, New York, 1995.
2. Piela, L., Nemethy, G., and Sheraga, H. A. *Biopolymers 26*, 1587, 1987.
3. Sankararamakrishnan, R., and Vishveshwara, S. *Biopolymers 30*, 287, 1990.
4. Dasgupta, S., and Bell, J.A. *Int. J. Pep. Protein Res. 41*, 499, 1993.
5. DeTar, D.F., and Luthra, N.P. *J. Am. Chem. Soc. 99*, 1232, 1977.
6. Cabrol, D., Brock, H., and Vasilescu, D. *Int. J. Quantum Chem. 6*, 365, 1979.
7. Schellman, J.A., and Neilson, E.B. In *Conformation of Biopolymers*, Ramachandrau, G.N., ed. Academic, London, 1976.
8. Brunne, R.M., van Gunsteren, W.F., Bruschweiler, R., and Ernst, R.R. *J. Am Chem. Soc. 115*, 4764, 1993.
9. van Gunstern, W.F., and Berendsen, H.J.C. *Groningen Molecular Simulations (GROMOS)*, Library Manual, Biomos. Groningen, 1987.
10. Kessler, H., Muller, A., and Pook, K.H. *Liebigs Ann. 903*, 1989.
11. Kessler, H., Bats, J.W., Lautz, J., and Muller, A. *Liebigs Ann. 913*, 1989.
12. Patel, D.J. *Biochemistry 12*, 667, 1973.
13. Madi, Z.L., Griesinger, C., and Ernst, R.R. *J. Am. Chem. Soc. 112*, 2008, 1990.
14. van Gunsteren, W.F., and Berendsen, H.J.C. *Mol. Simulations 1*, 173, 1988.
15. Shi, Y.Y., Wang, L., and van Gunsteren, W.F. *Mol. Simulations 1*, 369, 1988.
16. Hurley, J.H., Mason, D.A., and Matthews, B.W. *Biopolymers 32*, 1443, 1992.
17. Matthews, B.W., Nicholson, H., and Becktel, W.J. *Proc. Natl. Acad. Sci. USA 84*, 6663, 1987.

18. Nicholson, H., Soderlind, E., Trontrud, D., and Matthews, B.W. *Biopolymers 32*, 1431, 1992.
19. Summers, N.L., and Karplus, M. *J. Mol. Biol. 216*, 991, 1990.
20. Schimmel, P.R., and Flory, P.J. *J. Mol. Biol. 34*, 105, 1968.
21. Nicholson, H., Tronrud, D.E., Becktel, W.J., and Matthews, B.W. *Biopolymers 32*, 1431, 1992.
22. Alber, T., Bell, J.A., Dao-Pin, S., Nicholson, H., Wosniak, J.A., Cook, S., and Matthews, B.W. *Science 239*, 631, 1988.
23. O'Neil, K., and DeGrado, W.L. *Science 250*, 646, 1990.
24. Weiner, S.J., Kollman, P.A., Nguyen, D.T., and Case, D.A. *J. Comp. Chem. 7*, 230, 1986.
25. Sauer, U., Dao-Pin, S., and Matthews, B.W. *J. Biol. Chem. 267*, 2393, 1986.
26. Yun, R.H., Anderson, A., and Hermans, J. *Proteins 10*, 219, 1991.
27. Tronrud, D.E., Ten Eyck, L.F., and Matthews, B.W. *Acta Cryst. A 43*, 489, 1987.
28. Sapse, A.M., Mallah-Levy, L., Daniels, S.B., and Erickson, B.W. *J. Am. Chem. Soc. 109*, 3526, 1987.
29. Karle, I.L. *J. Am. Chem. Soc. 100*, 1286, 1978.
30. Matsuzaki, T., and Itaka, Y. *Acta Cryst. A 27*, 5007, 1971.
31. Mizushima, S., Shimanoucki, T., Tsuboi, M., Sugita, T., Kurosaki, K., Matagu, M., and Souda, S. *J. Am. Chem. Soc. 74*, 4639, 1952.
32. Pullman, B., Maigret, B., and Perahia, D. *Theor. Chim. Acta 18*, 44, 1970.
33. Ramek, M., Kerterer, A.M., Teppen, B. J., and Schafer, L. *THEOCHEM 352*, 59, 1995.
34. Zimmerman, S.S., Pottle, M.S., Nemethy, G., and Scheraga, H.A. *Macromolecules 1*, 10, 1977.
35. Kelterer, A.M., Ramek, M., Frey, R.F., Cao, M., and Schafer, L. *THEOCHEM 310*, 45, 1994.
36. Thomasson, K.A., and Applequist, J. *Biopolymers 30*, 437, 1990.
37. Matsuzaki, T., and Iitaka, Y. *Acta Cryst. B 27*, 507, 1971.
38. Bendetti, E., Christense, A., Gilon, C., Fuller, W., and Goodman, M. *Biopolymers 22*, 305, 1976.
39. Madison, V., Atreyi, M., Deber, C.M., and Blount, E.R. *J. Am. Chem. Soc. 96*, 6725, 1974.
40. Ramek, M., Keltere, A.M., and Nikolic, S. *Proceed. of the Sanibel Symp. (Int. Quantum Chem.)* 1997.

7
Taurine and Hypotaurine

Taurine (2-aminoethanesulfonic acid) is the end product of methionine and cysteine metabolism. It is found everywhere in the animal kingdom. In retina of vertebrates, including humans, it is the most abundant free amino acid, and it ranges from 10 to 52 µmol/g tissue (1).

In mammals, taurine is found in high concentrations in brain, retina, myocardium, liver, kidney, and muscles, as well as in platelets, lymphocytes, and cerebrospinal fluid (2–3).

Taurine is involved in osmotic regulation of cell volume in sharks, fishes, and marine bacteria (4). It is important as a bile acid conjugant in mammals (2), and it has been proposed (4) that it stabilizes sarcolemmal membranes in the heart, influencing thus the calcium flux.

Korang et al. (5) studied the levels of taurine, other amino acids, and related compounds in the plasma, vena cava, aorta, and the heart of rats after taurine administration. They present a pharmacokinetic study of the fate of taurine administered in blood and other tissues. They found that taurine administered by injections causes quite elevated plasma levels, as high as 70-fold at fifteen minutes. Later, these levels decrease and approach normal values after approximately four hours. Beta alanine and phospho-serine are also increased. Conversely, a number of amino acids, such as thre-onine, asparagine, glutamine, alanine, citrulline, and others are reduced. After half an hour, the level of taurine is doubled in vena cava and heart and tripled in the aorta. Effects on other amino acids in these tissues were also observed, and it was concluded (5) that pharmacological effects seen after taurine administration could be caused either by taurine level eleva-tion or by taurine-induced changes in other amino acids.

Kamisaki et al. (6) studied the effects of taurine on GABA release from synaptosomes of rat olfactory bulb. Superfusion of synaptosomes prepared from rat olfactory bulb show constant release of endogenous taurine. In addition, release of aspartate, glutamate, and GABA were also found (6).

Taurine release was calcium-independent (6). The perfusion with synap-tosomes with taurine in a 10 µM concentration inhibited the evoked release of GABA by 63% but did not change the basal release. Studies suggest that taurine may be released in large quantities form the nerve endings of the rat olfactory bulb, and it may regulate GABA release through the activation of presynaptic GABA autoreceptors (6).

Taurine is present in the retina of mammals, as mentioned before. Liebowitz et al. (1) studied the cat's retina from the point of view of taurine content and activity. The importance of taurine in cat's retina has been recognized (7–9). Without a taurine content in the diet, there is a severe degeneration of the photoreceptors, leading to blindness (1). As far as humans are concerned, the requirement for taurine in the diet has been established by Gegell et al. (10). It has been shown that taurine stimulates ATP-dependent calcium ion uptake at low concentrations and inhibits ATP-dependent calcium ion uptake when the calcium concentration is high (11–12). In order for taurine to demonstrate biological activity in the ATP-dependent calcium uptake system, there are structural requirements (11–12). Liebowitz et al. (1) examined the structure–activity relationships among a number of aminocycloalkanesulfonic acids, which are conformationally restricted analogues of taurine.

They present evidence that different cyclic analogues of taurine have different effects on ATP-dependent calcium ion uptake, with some analogues being inhibitors and others acting as stimulators (1). There is an inverse relationship in the effect of cyclic inhibitors on ATP-dependent calcium ion uptake and protein phosphorylation (1).

The taurine analogues examined by Liebowitz et al. are TAHS (trans-2-aminocyclohexanesulfonic acid), CAHS (cis-2-aminocyclohexanesulfonic acid), TAPS (trans-2-aminocyclopentanesulfonic acid), CAPS (cis-2-aminocyclopentanesulfonic acid).

The interatomic distances between the nitrogen and the sulfur atoms were estimated by the Dreiding stereomodels to be 4.0 A for TAPS and 2.7 A for CAPS. NMR NOE suggested a distance between vicinal protons of 2.7 A for TAPS and 2.3 A for CAPS, thus confirming trans and cis orientations. The interatomic distances in TAHS and CAHS were estimated to be 3.1 and 3.0 A, respectively. All the above distances correspond to the most stable conformations, as found by NMR spectroscopy (1).

In rat retinal tissues, Liebowitz et al. (1) report that TAPS is the strongest inhibitor of ATP-dependent calcium uptake. Taurine and CAHS are stimulators, while CAPS and TAHS are inhibitors. Conversely, TAPS and TAHS are stimulators of phosphate incorporation into rat retinal membrane preparations, taurine is somewhat of an inhibitor, and CAPS has no effect at small concentrations but becomes a stimulator at concentrations as high as 20 mM.

As pointed out by Liebowitz et al. (1), the fact that taurine stimulates ATP-dependent calcium ion uptake in retinal membrane is known, but the mechanism by which it does so is not clear. Definitely, ATP must play a role since in its absence taurine does not stimulate the calcium uptake. Moreover, the stimulation seems to be specific for taurine and similar compounds. Liebowitz et al. (1) postulate that rotationally restricted analogues of taurine with small molecular weights would have similar effects on calcium uptake. However, out of the examined analogues, only CAHS is a stimulator, the others being inhibitors.

The explanation of the effects of these compounds on the ATP-dependent calcium uptake and of the protein phosphorylation cannot be found only in terms of the interatomic distances between the nitrogen and sulfur atoms (1). A detailed structural investigation including electrostatic effects might provide information on the issue.

Taurine has been found to be the single most stimulatory amino acid among the amino acids responsible for the activity of an extract with low molecular weights of the antennular chemosensory system in *Panulirus argus* (spiny lobster) (13). Taurine has proved to be a stimulant in many crustacean studies (14–15). Indeed, taurine-sensitive receptors with response to very low thresholds are found on both the lateral and medial antennular filaments of *Panulirus argus*. By examining the molecular specificity of the receptors with respect to a comparison of the stimulatory capacity of taurine with the stimulatory capacity of taurine analogues and other structurally related compounds, Fuzessery et al. (13) shed light on the stimulatory mechanism. They found that the antennular taurine receptors on the spiny lobster feature a narrow specificity similar to the taurine endoreceptors of other organisms.

In each experiment performed by Fuzessery et al. (13), a search was conducted for single taurine-sensitive neurons during stimulation with taurine.

The stimulatory capacity of taurine was compared with that of five sulfonic acids and three analogues (13). It was found that only taurine and its carboxylic and sulfinic analogues, beta alanine, and hypotaurine, stimulated all the receptors. Fuzessery et al. (13) conclude that:

- The most effective compounds are those with one terminal basic group and one terminal acidic group, separated by two carbon atoms. Taurine belongs to this group, as well as hypotaurine and beta alanine, with the last two less effective than taurine. The phosphonic acid, also meeting the structural requirements, is much less stimulatory than taurine. It also elicits a response from only one of the tested receptors.
- The stimulatory activity decreases with the number of carbons that separate the acidic group and the basic group, as long as the number is two or higher. For instance, 2-aminobutyric acid and 3-aminobutyric acid are less effective than their isomer, GABA.
- Compounds in which the acidic and the basic groups are separated by only one carbon atom are much less effective.
- Taurine derivatives without the basic group are much less active.
- A neutral side chain added decreases the activity of a compound. For example, beta-aminoisobutyric acid is much less effective than beta alanine.

- Compounds with an alpha amino group besides the terminal amino group are inactive. As an example, DABA is inactive.
- Compounds in which the basic terminal group is a guanidinium moiety rather than an amine are much less effective.
- The activity disappears when the carboxyl of an active acid such as beta alanine is involved in a peptidic bond.

These results, coupled with quantitative studies, indicate that the antennules of *Panulirus argus* indeed possess taurine-sensitive receptors with a narrow and consistent specificity (16). The same conclusion was reached by Shepheard (16) on another crustacean, *Homarus americanus*. In that study, out of 43 amino acids tested for stimulatory activity, only taurine and beta alanine showed activity.

In *Panulirus argus*, the taurine activity is mimicked by hypotaurine and beta alanine. These compounds are similar in activity in other instances, such as suppressing induced heart seizures in dogs (17). GABA is also active in this respect, but glycine and alpha alanine are not. It also has been found that in an active transport system in human blood platelets, taurine uptake is inhibited competitively by beta alanine and hypotaurine but not by the phosphonic acid analogue (18).

Taurine is the most abundant beta amino acid in most marine animals (13). As shown by Awapara (19), "taurine exists uncombined and distributed throughout the animal kingdom in a manner almost unparalleled by any known small organic molecules."

Kontro and Oja (20) investigated the effects of structural analogues on releases of taurine and GABA, either spontaneous or potassium-stimulated. It is known that GABA and taurine cross the brain cell membranes by a common uptake system (21–22). This system contains two components: high-affinity and low-affinity. Kontro and Oja (20) attempted to infer whether the carrier-mediated exit of GABA and taurine can occur by the common reversed transport in both spontaneous and depolarization-induced release.

They found that the potentiation of GABA and taurine release by homoexchange depends on the concentration. The common transport system has a much higher affinity for GABA than for taurine (23).

Kontro and Oja (23) showed that taurine release was potentiated by kainic and glutamic acid. This indicate that excitatory amino acids influence taurine release. Kainic-induced epilepsy causes an increase of taurine in the extracellular region of the hippocampus, in vivo (24).

The potassium-stimulated taurine release was suppressed by exogenous taurine and its analogues in a concentration-dependent way (23).

Debler and Lajtha (25) studied the high-affinity transport of several amino acids, including taurine. They point out that taurine's role as a neutrotransmitter has been suggested, but only in a speculative way. However, the transport mechanisms of GABA and taurine show some overlap. Debler and Lajtha (25) demonstrated the existence of three separate high-affinity transport systems of GABA, taurine, and glycine. As mentioned before, there is overlap between the GABA and taurine transport, but not between taurine and glycine transport. Hypotaurine inhibits strongly GABA transport, indicating overlap with taurine (26). Taurine appears to feature a high-affinity uptake, less specific than that of GABA.

Taurine has been found to exhibit an anticonvulsant effect (27). Huang et al. (27) observed the anticonvulsant effect of taurine with or without sodium phenobarbital. This effect was studied in mouse convulsion model induced with pentetrazol. The influence on GABA, taurine, and glutamate levels in the brains was determined. Taurine was found to prolong the latent period and reduce mortality (27). When sodium phenobarbital was added to taurine, the effect was increased. Taurine was found to increase the levels of GABA and taurine in the brain of the mice (27). Huang et al. (27) suggest that the anticonvulsant effect of taurine may be associated with its modulating the functions of the central glutamate–GABA axes.

Franconi et al. (28) compared taurine and several of its analogues with respect to their action on tension in guinea pig ventricular strips. Taurine-like activity had a strict requirement for the presence of two carbon atoms between the sulfonic and amino groups. This observation might lead to information about the size and shape of the binding site on the taurine receptor. That study also found that if the aliphatic chain is replaced by an aromatic ring, activity is present only when the sulfonic and the amino groups are bound to adjacent carbons on the ring. A possible explanation (28) is that taurine binds in a planar conformation.

Experimental results (29) show that taurine crosses the blood–brain barrier very slowly, while the more lipophilic pivaloyltaurine does so readily. The conformational structure of taurine might have an influence on its ability to penetrate the blood–brain barrier.

In order to shed additional light on taurine's structure and mechanism of action in various respects, Fugler-Domenico et al. (30) applied ab initio calculations to the study of taurine and hypotaurine.

Three conformations of taurine and hypotaurine were examined: an extended conformation, a partially folded conformation, and a cyclic conformation. The energies and optimum geometries of the conformers were obtained via ab initio calculations, at Hartree–Fock level, using the 6-31G basis set. In addition, single-point 6-31G* energies were calculated at the 6-31G obtained optimum geometry. Each system was geometry optimized by allowing all the molecular parameters of the molecule to relax, subject to the following constraints: The NH and CH bond lengths as well as the HCC and the HNC angles were kept equal. Also, the OSC angles and the SO bond lengths were kept equal. The gradient method was used. In addition to the neutral species, cations of taurine and hypotaurine, obtained by adding a proton to the amino group, were also investigated.

The geometry optimization was performed making sure that the second-derivative matrix contains only positive elements, so the conformation represents a real minimum.

Figures 7.1 and 7.2 show the various conformations of taurine and hypotaurine, respectively. Fugler-Domenico et al. (30) found that the most stable conformation for taurine is the cyclic conformation, depicted in Figure 7.1c. The cycle is formed due to the presence of a hydrogen bond between the hydrogen of the sulfonic group and the nitrogen of the amino group. This result is in agreement with the conclusions of Franconi (28) and Liebowitz and Lombardini (31), who suggest a gauche conformation around C1C2 for taurine. Such a conformation could feature a hydrogen bond similar to the one shown in Figure 7.1c. A conformation featuring a hydrogen bond between one of the hydrogens of the amino group and one of the oxygens of the sulfonic group, shown in Figure 7.1d, was found to be high in energy. The extended conformer of taurine, shown in Figure 7.1a, is higher in energy than the cyclic conformation by only 3 kcal/mole. Both are thus likely to coexist in vivo. The ab initio calculations performed in this study do not take into consideration the solvent effect, and as such, the zwitterion that features a protonated amino group and an unprotonated sulfonic group is found to be less stable. However, it is possible that in a hydrated medium the zwitterion would be more stable. The partially folded conformer is higher in energy than the extended one. The most stable cation for taurine, shown in Figure 7.1e, is cyclic.

In hypotaurine, since the third oxygen is absent, the nitrogen can position itself close to the sulfur. The most stable conformation is shown in Figure 7.2b, and it is the partially folded conformation. Next in stability is the extended conformation, 7.2a, higher in energy by only 1 kcal/mole. The conformer that features a hydrogen bond between the hydrogen on the sulfinic group and the amine's nitrogen (Figure 7.2c) is higher in energy than that of Figure 7.2b by 4 kcal/mole.

As shown in discussions of GABA, the ability of these types of molecules to penetrate the blood–brain barrier is related to their lipophilicity, which increases when the most stable conformer is cyclic. Another factor that influences the capability to cross the barrier is the proton affinity of the amine group and of the acid group. A high proton affinity at NH_2 facilitates the formation of NH_3^+, while a low proton affinity at the acidic end facilitates the formation of an anion. It was found (28) that the proton affinity of the amine group of taurine is smaller than that of GABA and GABA imine by 8.3 kcal/mole. The proton affinity of the amine group of hypotaurine is similar to that of GABA. As far as the acidic groups are concerned, at 6-31G calculational level, the proton affinity of sulfonate is smaller by 38 kcal/mole than that of the carboxyl group (30). These differences might account for the ability of taurine to cross the blood–brain barrier, albeit slowly.

As found by Debler and Lajtha (25), the synaptosomal uptake of taurine is very inhibited by beta alanine and hypotaurine and somewhat less by

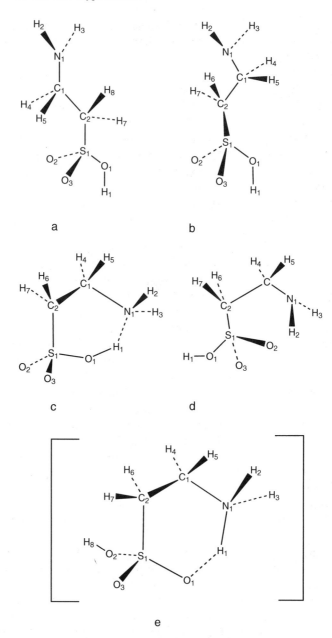

FIGURE 7.1 a. Extended conformer of taurine. b. Partially folded conformer of taurine. c. Cyclic conformer of taurine with hydrogen bond between H_1 and N_1. d. Cyclic conformer of taurine with hydrogen bond between O_2 and H_2. e. Cyclic taurine cation with H at N_1.

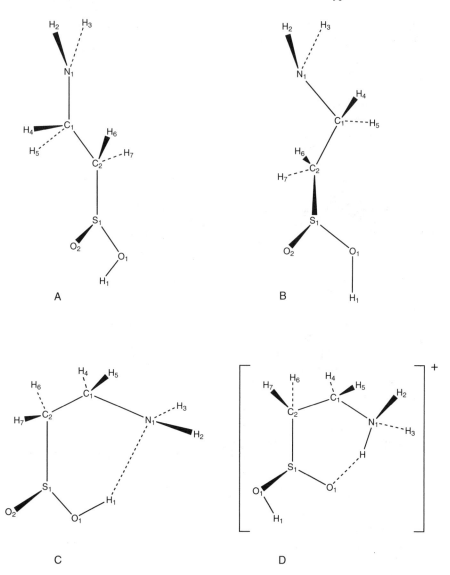

FIGURE 7.2 a. Extended conformer of hypotaurine. b. Partially folded conformer of hypotaurine. c. Cyclic conformer of hypotaurine. d. Cyclic hypotaurine cation with H on N_1.

GABA and proline. If it is hypothesized that the unhydrated structures are recognized by the receptor binding sites, a partially folded conformation of GABA, DABA, and hypotaurine would fit into the GABA binding site. In the taurine-receptor interaction the decisive factor might be related to the distance between the amino group and the acidic group. This distance is very similar in taurine and hypotaurine in their most stable conformations (30), as well as in beta alanine.

References

1. Lebowitz, S.M., Lombardini, J.B., and Allen, C.I. *Biochemical Pharmacology 37*, 1303, 1988.
2. Wright, C.E., Tallan, H.H., and Lin, Y.Y. *Ann. Rev. Biochem. 55*, 427, 1986.
3. Gaull, G.E. *Acta Paeditr. Scamd. Suppl. 296*, 1982.
4. Chesney, R. *Adv. Ped. 32*, 2, 1985.
5. Koran, K., Milakofsky, L., Hare, T.A., Hofford, J.M., and Vogel, W.H. *Pharmacology 52*, 263, 1996.
6. Kamisaki, Y., Wada, K., Nakamoto, K., and Itoh, T. *Amino Acids 10*, 49, 1996.
7. Hayes, K.C., Carey, R.E., and Schmidt, S.Y. *Science 188*, 949, 1975.
8. Schmidt, S.Y., Berson, E.L., and Hayes, K.C. *Invest. Ophtal. 15*, 47, 1976.
9. Berson, E.L., Hayes, K.C., Rabin, A.R., Schmidt, S.Y., and Watson, G. *Invest. Ophtal. 15*, 52, 1976.
10. Geggel, H.S., Ament, M.E., Heckenlively, J.R., Martin, D.A., and Kopple, J.D. *New Eng. J. Med. 312*, 142, 1985.
11. Lombardini, J.B. *J. Neurochem. 40*, 402, 1983.
12. Pasantes-Morales, H., and Ordonez, A. *Neurochem. Res. 7*, 317, 1982.
13. Fuzessery, Z.M., Carr, W.E.S., and Ache, B.W. *Biol. Bull. 154*, 226, 1978.
14. Allison, P., and Dorsett, D.A. *Mar. Behav. Physiol. 4*, 205, 1977.
15. Carr, W.E.S., and Gurin, S. *Biol. Bull. 148*, 380, 1975.
16. Shepheard, P. *Mar. Behav. Physiol. 2*, 261, 1974.
17. Barbeau, A., Tsukada, Y., and Inoue, N. In *Taurine*, Huxtable, R. and Barbeau, A. (eds) Raven Press, New York, 1976.
18. Grant, Z.N., and Nauss, C.B. In *Taurine*, Huxtable, R. and Barbeau, A. (eds) Raven Press, New York, 1976.
19. Awapara, J., Grant, Z.N., and Nauss, C.B. In *Taurine*, Huxtable, R. and Barbeau, A. (eds) Raven Press, New York, 1976.
20. Kontro, P., and Oja, S.S. *Neurochem. Res. 12*, 475, 1987.
21. Martin, D.L., and Shain, W. *J. Biol. Chem. 254*, 7076, 1979.
22. Kontro, P., and Oja, S.S. *Neurochem. Res. 6*, 1179, 1981.
23. Kontro, P., and Oja, S.S. *Neurochem. Res. 8*, 1377, 1983.
24. Lehmann, A., Hagberg, H., Jacobson, J., and Hamberger, A. *Brain Res. 359*, 147, 1985.
25. Debler, E.A., and Lajtha, A. *J. Neurochem. 48*, 1851, 1987.
26. Kontro, P. In *Amino Acid Neurotransmitters*, eds. DeFeudis, F.V. and Mandel, P. Raven Press, New York, 1981.
27. Huang, F., Liu, B., Chen, Y., and Lin, H. *Guangdong Yaoxueyuan Xuebao 11*, 220, 1995.
28. Franconi, F., Failli, P., Stendardi, I., Matucci, R., Bernadini, F., Baccarro, C., and Giotti, A. *Eur. J. Pharm. 124*, 129, 1986.
29. Oja, S.S., Kontro, P., Linden, I.B., and Gothoni, G. *Eur. J. Pharm. 87*, 191, 1983.
30. Fugler-Domenico, L., Russell, C.S., and Sapse, A.M. *Biogenic Amines 6*, 289, 1989.
31. Liebowitz, S.M., and Lombardini, J.B. *Suppl. Invst. Ophtal. Vis. Sci. 25*, 260, 1984.

8
Ab Initio Calculations Related to Glucagon

Glucagon is a polypeptide (29 residues) hormone that regulates gluconeo-genesis and glycogenesis in the liver. Its effects are mediated by cAMP (cyclic adenosine monophosphate), which is synthesized in a reaction cat-alyzed by membrane-bound adenylate cyclase. As discovered by Earl Sutherland in the 1950s, glucagon binds to receptors on the surface of liver cells, and there they trigger the formation of cAMP. While the hormone itself is thought not to penetrate the cell, cAMP mediates the intracellular effects, serving thus as a second messenger to the hormone, which is the first messenger.

Glucagon triggers the breakdown of glycogen. The signal transduction pathway from hormone to the degradation of glycogen can be now explained in molecular terms in the following way (1):

- Glucagon binds to seven-helix receptors in the membrane of the cells and activates the stimulatory G protein.
- The GTP form of G activates adenylate cyclase, which catalyzes the for-mation of cAMP from ATP.
- Protein kinase A is activated.
- Protein kinase A phosphorylates glycogen synthase and deactivates it.
- The glucagon effect is highly amplified by the cAMP cascade.

In insulin-deficient diabetes glucagon is present in the blood in large quantities, showing that it plays a role in the pathogenesis of diabetes mel-litus (2). In patients suffering from this disease the level of insulin is too low, and the level of glucagon is too high. This abnormality promotes glyco-gen breakdown, with an excessive amount of glucose released into the blood.

The binding of the glucagon to its receptors on the cell's surface is highly specific. Other structurally related hormones, such as secretin and vasoac-tive intestinal peptide, do not bind to the glucagon's receptors (2). More-over, small changes in the amino acids' sequence can produce large changes in biological activity. These results show that the binding of the glucagon hormone to its receptors is very rigorous and very specific. Accordingly, the three-dimensional arrangement of the peptide has to be elucidated in order to understand its binding to the receptor. The X-ray structure of glucagon has been determined (3), but its solution structure is not yet completely known. Chou and Fasman (4,5) have performed probability calculations that have formed the basis of a model proposed by Korn and Ottensmayer (13) for the solution structure of glucagon.

In order better to understand the activity of glucagon and its binding to the receptor, Unson et al. (2,6) have synthesized a series of glucagon antagonists and studied their binding and their activities. These antagonists were designed as synthesizing replacement analogues with altered peptide chain conformation. Some of these derivatives (6) were completely inert toward the adenylate cyclase activation in the liver plasma membrane. However, they showed weak binding affinity and could act as glucagon inhibitors.

Another approach to the search for glucagon antagonists taken by Unson et al. (6) made use of the fact that secretin, a structural analogue of glucagon that does not bind to the glucagon receptors on the cells of liver, may have evolved from glucagon through a series of intermediates that retained binding ability but lost the ability to activate adenylate cyclase. It is thus possible to find hybrids of secretin and glucagon that might feature the desired properties. Examining the sequences of secretin and glucagon, one observes that in the first 12 residues are three changes: In glucagon the residues 3, 9, and 12 are Glutamine (Gln), Aspartic Acid (Asp), and Lysine (Lys). In secretin, these are changed to Asp, Glutamic Acid (Glu), and Arginine (Arg), respectively. The synthetic hybrids with glucagon with Asp3, Glu9, and Arg12, as well as with Asp3 and Glu9 only, were totally inactive (6). However, even though they did not produce cAMP, they still retained 2% of glucagon's binding activity. It was concluded that the change responsible for this effect was Asp9 to Glu9. In addition, Unson et al. (6) showed that an introduction of a C terminal amide produces a fully inactive peptide with high receptor activity and a good inhibition index.

It has been shown (7,8) that Asp3, Glu9, Arg12 glucagon, and Asp3, Glu9 glucagon bind to liver membranes only 2% as tightly as glucagon, but their activity is even less (less than 0.001% that of glucagon). However, the Asp3 glucagon, which is a single-replacement analogue, bonded slightly but showed more of an agonist-like character. Its maximum adenylate cyclase activity was 17%. These results also indicate that the site responsible primarily for the lack of activity is Glu9.

Unson et al. (6) follow the changes in membrane binding of the synthesized products by competition with ^{125}I-labeled glucagon, and in adenylate cyclase activation via cAMP measurements. They define the inhibition index as the ratio of the inhibitor to agonist when the response is reduced by 50% of the response to the agonist when the inhibitor is not present. The antagonism is also represented as the pA2 value, which represents the negative value of the logarithm of the inhibitor concentration that makes one unit of agonist exhibit the response obtained from 0.5 units of agonist.

The replacement of Asp9 by Glu9 or the deletion of His1 in glucagon reduced binding to 6% or 8% (6) and produced agonists with 24% or 36% responses, respectively, even at high concentrations. However, when both changes are made, the product so obtained features 11% binding but no activity (6). If a third change was introduced, consisting in the replacing of the normal carboxyl group at one end by a terminal primary amide, the binding affinity was increased, especially when it was in conjunction with the His1 deletion. However, if des-His1 glucagon amide still had some low potency, the des-His1 Glu9 glucagon amide had high binding affinity but no adenylate cyclase activity. It was a pure antagonist, with an inhibition index as low as 12.

Unson et al. (6) state that making a pure antagonist of a peptide hormone can provide an important tool for the study of the mechanism of action of the hormone. In addition, some antagonists can be of therapeutic value in diseases where the activity of the natural endogenous hormone is too high. It has been shown (9) that N-trinitrophenyl (12-homoarginine) glucagon prepared from natural glucagon binds to the glucagon receptor but does not produce cAMP. However, it activates another binding system and produces inositol triphosphate and Ca^{2+}. Furthermore, it demonstrated the existence of a second glucagon receptor. It has also been shown (10) that the acinar cells of the pancreas have two secretin receptors, one producing cAMP and the other producing inositol triphosphate and Ca^{2+}.

Unson et al. (6) searched for potent glucagon inhibitors using the following guidelines:

- Changes in the conformation of the 19–27 region will possibly be related to receptor binding and activity.
- Secretin and glucagon hybrids may bind to the receptors but do not transduce the signal for the production of cAMP.
- The deletion of the N-terminal histidine reduces binding and activity in different ways.
- The addition of a C-terminal amide increases the helical dipole, which could play a role in the binding to the receptor (11).

In a previous work, Lu et al. (12) have shown that the removal of histidine 1 from Tyr22 glucagon produces a molecule that somewhat retains the membrane binding capacity but no adenine cyclase activation.

The introduction of C-terminal amide groups into the glucagon analogues was based on the fact that many biologically active peptides that terminate in amides are more active than those that terminate in carboxyls (11). It has been shown via circular dichroism methods (6) that the introduction of the amide increases the proportion of alpha helices and decreases the beta-sheet structure. Such a change can lead to a greater flexibility of the glucagon molecule and therefore to a better binding to the receptor. However, the changes in hormone signal transduction are not related to the amide but to the removal of His1 and to the replacement of Asp9 by Glu (6).

Unson et al. (2) have synthesized additional analogues of glucagon containing the des-His1 and Glu9 changes and the COOH-amide substitution. The antagonists so obtained feature inhibition indices from 10 to 300.

Fifteen purified analogues of glucagon without His1, with Glu9, and with amide instead of the terminal carboxyl, as well as four purified analogues also containing replacements in the 2–5 positions were evaluated for their ability to bind to the glucagon receptor and to produce cAMP in the hepatocyte plasma membranes. Amino acid substitutions in the 12th position were made in order to evaluate the effect of this change on the glucagon's properties. It was thought (2) that a salt bridge formed by the Asp9 residue or its replacement, the Glu9 residue with the Lys12, plays a role in the binding to the receptor and that it is responsible for the retention of the binding when Asp9 is replaced by Glu. Indeed, the replacement of the lysine by a norleucine residue, which is hydrophobic and uncharged, caused a 20-fold decrease in the binding activity as compared to glucagon. This reduction in binding increased the inhibition index, requiring 300 times as much of the analogue to reduce an agonist response to 50%. Homoarginine analogues, though, proved to be partial agonists with binding ability. In contrast, the binding affinity is greatly reduced when lysine is replaced by a glutamic residue featuring a negative charge (2).

The adenylate cyclase activity of all these analogues is greatly reduced as compared to glucagon (2).

By placing a negatively charged residue at the positive end and a positively charged residue at the negative end, the alpha-helix structure is accentuated (2). Some of these changes resulted in analogues that lost all their binding ability.

Another approach used by Unson et al. (2) in the study of the binding ability and adenine cyclase activity of glucagon was to synthesize smaller fragments of the hormone, such as deleting the 1 to 6 residues in the Glu9 glucagon. They found that in this case the analogue retains only 0.33% of the binding affinity. If in addition residues 7 to 8 are deleted, all binding affinity is lost. However, if only the 1 to 4 residues are removed, the binding affinity is 27% of that of the native hormone. It has been shown (13) that the 2 to 5 residues are involved in the formation of a beta turn, which may play some role in the binding to the receptor.

A point of great interest was the fact that the cyclase activity was decreased significantly when Asp9 was replaced by Glu9. A negative charge is conserved in this replacement, and 40% of the binding is conserved, probably due to the conserving of the salt bridge between the negative residue and the positive residue of Lys12. As mentioned before, if the Lys12 residue is replaced by a neutral or negative residue, the binding is significantly decreased. However, even if the Lys12 residue is present, if Asp9 is replaced by Glu9, the activity is greatly reduced. It would be of great informative value to find an explanation for this fact.

It is believed that the receptors of the glucagon peptide hormone family that are coupled to adenylate cyclase via a guanine nucleotide regulatory protein are similar (2). It was shown (14) that an aspartic acid residue is necessary for agonist binding in one of the transmembrane helices. This residue might serve as counterion to the amino group of the catecholamine for the beta-adrenergic receptor. Similarly, an aspartic residue in the glucagon receptor and in the glucagon peptide hormone family of receptor proteins might be a counterion to the lysine residue in glucagon or in the growth hormone releasing factor, as well as to the Arg12 in secretin and vasoactive intestinal peptide (2). Indeed, position 12 seems to be a common site of attachment to a corresponding complementary position in the receptor. Unson et al. (2) propose that the salt bridge is not formed between position 9 and 12 of glucagon but between these residues and complementary sites on the receptor protein.

The analogues synthesized by Unson et al. (2) show, through circular dichroism methods, an increase in the beta sheet percentage and a decrease in the beta turns. These changes seem to be correlated to both the changes in binding and in transduction activity. It appears that a positive charge at the 18 position is essential for both the receptor binding and the transduction of the signal. It also appears that correlation between activity and structure is more evident in concentrated phases than in dilute solutions because the molecules can feature a greater flexibility in the latter.

The fact that residues 1 to 8 are essential for binding is explained in the following way (2): Glucagon is degraded by a liver aminodipeptidase. This enzyme cleaves dipeptides from the amino terminus, leaving a piece of the peptide that binds to the receptor but does not show adenylate cyclase stimulation. As such, the receptor is sensitive to even small conformational changes at the amino terminus. As shown by Unson et al. (2), their data indicate that the amino terminus does not have as rigid a requirement for a turn, and it is possible that it assumes an extended or random conformation. This hypothesis is supported by NMR studies (15).

In order better to understand the reason for the changes introduced in the binding and activity of glucagon by the replacing of Asp9 by a glutamic residue, as shown in Chapter 5, the aspartic acid and the glutamic acid have been geometry optimized. In addition (ref. 2, chapter 5), the two molecules have been superimposed, using the Insight II computer program.

The goal of this work is to dissociate structural and conformational properties of the hormone that are important for binding, that is, for the recognition message, from those necessary for transduction, that is, the biological activity message. Since the experimental results have demonstrated that the replacement of Asp9 by Glu does not affect the binding too much but leads to a product that does not have adenylate cyclase activity, it was thought that the reason might lie in the difference in the position of the carboxyl group in the two amino acids.

If the aspartic and the glutamic acids in their most stable conformations as determined by ab initio calculations are superimposed, one can see (ref.

2, chapter 5) that the position of the carboxyl group differs significantly. As such, at the site where in the aspartic residue there is a negative charge, in the glutamic one there is a slightly positive CH_2. It is probable that the lysine or arginine of the receptor that provides the positive charge binding to the negative charge in the glucagon at the 9 position is flexible enough to bind also to the glutamic residue. However, the requirements for the transduction signal are probably more rigid, and the presence of the negative charge at the exact position present in the Asp9 species is crucial. In addition, Unson et al. (16) have demonstrated that the protonable imidazole group of histidine (His1) is required for interaction with Asp9. Ser16 was identified as an important residue for the expression of a full agonist response (17).

The involvement of histine, aspartic acid, and serine residues in an active intermediate is similar to the situation in serine protease. As such, it can be thought that Asp9 is part of a putative analytic triad, the other two amino acids being histidine and serine. When bound to the receptor, the glucagon acquires enzyme activity, and there is the formation of a charge relay network, containing the His, Asp, and Ser residues. This network takes part in the catalyzed hydrolysis of an amide bond in the receptor, with the carboxylate group of Asp9 stabilizing the positive imidazole ring of His1. Ser16 is probably involved in a nucleophilic attack on the receptor. Thus, the glucagon–receptor complex acquires serine protease-like activity. This proteolytic event might trigger the transduction. This hypothesis is supported by the fact that the serine-protease inhibitor 4-aminophenylmethanesulphonyl fluoride completely suppresses cAMP production by glucagon-stimulated liver membranes (18). It also may explain the antagonism in the case of des-His1-Glu9 glucagon amide and analogues with replacement in the 9 position. An antagonist may stabilize a conformation that either does not bring His, Asp, and Ser into the proper orientation or has the required conformation but not the required residue. In this case, replacing Asp by Glu should not completely destroy the activity since catalytic triads His-Glu-Ser are known (19). However, des-His1-Glu9 glucagon is a pure antagonist that is insensitive to GTP concentration (20), suggesting that the receptor moiety bound to this analogue is incapable of interacting with the G protein. It can thus be concluded that the Glu9 glucagon has a perturbed charge relay scheme and receptor autolysis does not occur. The receptor after autolysis is probably necessary to react with the G protein, which in turn activates the adenylate cyclase.

It is concluded (ref. 2, chapter 5) that replacing Asp9 by Glu destabilizes the protein and induces conformational changes. It changes the position of the charges on the side chain and produces an alteration of the charge relay mechanism.

The calculations described above are performed in gas phase. Also, only the neutral amino acids were considered, even though at physiological pH aspartic and glutamic acids exist in anionic form. Therefore, a subsequent work performed calculations pertaining to two tripeptides, Ser-Asp-Tyr and Ser-Glu-Tyr. The first is part of glucagon, namely, the 8, 9, and 10 residues, while the second is part of the glucagon analogue in which Asp9 was replaced by Glu. The geometry optimization of these tripeptides can shed some light on the glucagon's structure in solution. Indeed, even though the X-ray crystal structure of the hormone has been determined (3), its liquid phase structure is not yet known. Based on Chou and Fassman's probability calculations (4), Korn and Ottensmayer (13) have proposed a working model for glucagon in solution.

Sapse et al. (see Appendix) performed Hartree–Fock calculations using the STO-3G, the 3-21G, the 6-31G, and the 6-31G* basis sets, as implemented by the Gaussian-92 and the Gaussian-94 computer programs, on the neutral tripeptides and on the tripeptides featuring an unprotonated aspartic or glutamic residue, which thus become negatively charged. In addition, the solvent effect is examined by calculating the energy of interaction between the solvent treated as a continuum and each of a number of chosen conformers, in order to assess whether there are changes in the order of stability due to the solvent.

The tripeptidic fragment of glucagon that contains the aspartic residue in the middle was extracted via molecular modeling from the crystal structure of glucagon (21). The ends were blocked with one formyl group at one end and an amino group at the other. The fragment was geometry optimized with the STO-3G basis set. The aspartic residue was consequently replaced by a glutamic residue, and the geometry of the tripeptide Ser-Glu-Tyr was optimized. The angles particularly examined were the torsion angles characterizing the peptide bonds between serine and the aspartic or glutamic acid residues, and between the latter and tyrosine. These angles (OCCN) are called, respectively, alpha and beta.

For the neutral (protonated) tripeptides, it was found that the most stable conformer of Ser-Asp-Tyr feature an alpha angle of 158° and a beta angle of 241.2°. The Ser-Glu-Tyr tripeptide features an alpha angle of 203° and beta angle of 243°. It can thus be seen that the beta angles are quite similar, practically equal, while the alpha angles are somewhat different. As in the case of the separate amino acids, when the tripeptides are superimposed it can be seen that the positions of the carboxyl groups on the middle residues differ. According to the experimental results discussed above, this difference does not affect greatly the binding to the receptor, but it affects significantly the adenylate cyclase activity.

Several other conformers of Ser-Asp-Tyr and Ser-Glu-Tyr were examined via single-point calculations of the energy at different values of alpha and beta. It was thus found that keeping alpha at its optimized value and assigning beta values of 0° or 180° increases the energy by 6–7 kcal/mole. For these values of beta and a value of alpha of 268°, the energy increases by 10 kcal/mole. For both tripeptides, if beta takes a value of 60°, the energy increases greatly.

The solvent effect on the stability of various conformers was examined by two methods: the reaction field method, included in the Gaussian-92 and Gaussian-94 computer programs, and the AMSOL method.

The first method implements the Onsager reaction field model in which the solvent is considered a continuum characterized by its dielectric constant. The solute occupies a spherical cavity of radius r. The electric field due to the solvent's dipole interacts with the solute's dipole. The Hamiltonian of the system contains a term due to solvation, which describes the coupling between the molecular dipole operator and the reaction field, which is a function of the dielectric constant and the cavity radius. The cavity radius was determined by measuring the distance between the most widely separated atoms of the tripeptide, dividing the result by two, and adding the Van der Waals radius of the hydrogen atom. Since the tripeptide is not spherical, an error is introduced. This error makes the cavity larger than it really is, underevaluating thus the interaction energy. This error is similar, though, for the different conformers, and the comparisons are valid.

In addition, for the optimized conformers and for some of the other conformers considered, the interaction energy was estimated with the AMSOL-AM1 method (22). AMSOL combines semiempirical quantum-chemical calculations with continuum electrostatics and a surface-dependent dispersion, while AM1 gives the gas-phase semiempirical energy. The solvation energy was obtained as the difference between the AMSOL and the AM1 energies.

With both methods it was found that the glutamic-containing peptide features a larger solvation energy than the aspartic one. This can be explained by the fact that in the first, the carboxyl group is closer to the wall of the cavity.

For the neutral species, the solvation energy, as determined by either method, can at no point supersede the differences in energy between the different conformers. It may be concluded that the most stable conformers in gas phase are also the most stable in solution.

For the anionic species, it was found that the STO-3G basis set is inadequate for the description of the system, especially for the Ser-Asp-Tyr tripeptide. Indeed, geometry optimization using the STO-3G basis set results in the migration of the proton from the peptidic bond to the carboxylate ion. To verify that this result is wrong as a model for the system, the complex formed by the formate ion and formamide was geometry optimized, and indeed, with the STO-3G basis set, the proton migrates from formamide to the formate ion. However, when double-zeta basis sets are used (3-21G and 6-31G*), the proton stays on formamide. Consequently, the tripeptides were reoptimized with the 3-21G basis set, and no migration of the proton occurred. In addition, single-point calculations were performed using the 6-31G* basis set. To evaluate possible effects of the correlation energy, the MP2/6-31G*//HF/3-21G method was applied.

For the Ser-Asp-Tyr, the angle alpha was found to feature a value of 161.25° and the beta angle a value of −122.93°. Conformers with alpha of 90° are much higher in energy. Conformers within a 42 kcal/mole distance in energy from the optimized one have either the same alpha as the optimized one and betas of 0°, 90°, or 180°, or alphas of 0° and optimized beta or beta of 90°.

One observes that the 6-31G* and MP2/6-31G* calculations predict the same order of stability of various conformers as the 3-21G basis set calculations.

The Ser-Glu-Tyr peptide features an optimized alpha of 115.78° and an optimized beta of 168.21°. However, a conformer with the same alpha but a beta of −90°, a value similar to the −122° of Ser-Asp-Tyr, features an energy higher than the minimum by only 4.4 kcal/mole.

The superposition of the most stable conformers of the anionic tripeptides show more of a difference than for the protonated species.

Calculating the solvation energy with the reaction field method retains the gas-phase optimized conformation as the most stable. The same is true for the Ser-Asp-Tyr peptide when AMSOL calculations are performed. However, for the Ser-Glu-Tyr peptide, the AMSOL method coupled with higher-level ab initio calculations predicts as the most stable the conformer with alpha of 180° and beta of −90°, a conformer more similar to the Ser-Asp-Tyr tripeptide. One of the differences between the reaction field and the AMSOL method is that the reaction field method uses a sphere as a cavity, while the AMSOL method uses cavities appropriate to the shape of the solute molecule. A value of 180° for alpha affords better contact between solvent and solute in the AMSOL method than in the reaction field method. Therefore, the difference in activity for glucagon and Glu9 glucagon cannot be explained in terms of different conformations in solution but rather by the shifting of the carboxyl position.

To investigate the glucagon activity further, Carruthers et al. (23) have designed and synthesized a gene for the rat glucagon receptor. Their work is based on the fact that the isolation of the glucagon receptor cDNA clones from rat and human liver has confirmed that the receptor belongs to the superfamily of seven-transmembrane domain G protein-coupled receptors (24). It is thought that the hormone-binding site consists of a part from the large extracellular domain of the receptor, which includes the N—terminal tail and the loops connecting the transmembrane helices (23). The transmembrane signaling, though, must involve communication between the extracellular and the intracellular domain where the G proteins are activated by the receptor. This communication is mediated by the ligand to the receptor.

COS cells that express the synthetic receptor gene and purified COS cell membranes displayed high binding affinity toward glucagon and appropriate hormone specificity. In addition, transfected COS cells showed elevated cAMP levels in response to glucagon (23). However, site-directed mutant glucagon receptors with Asp64 replaced were expressed at normal levels in COS cells, but did not bind glucagon, indicating that Asp 64 may be crucial for the glucagon-receptor binding.

The synthetic gene built by Carruthers et al. (23) consisted of a nucleotide sequence that encodes the proper amino acid sequence of the glucagon receptor but contained a large number of restriction endonuclease recognition sites. High-level expression of the synthetic rat glucagon receptor gene in a vector where transcription was controlled by the human adenovirus major-late promoter was obtained in transiently transfected COS cells (23). In agreement with the results of Iwanij and Vincent (25), a receptor dimer was found.

The fact that the substitution of Asp64 in the receptor prevented the binding of glucagon indicates a direct interaction between Asp64 and glucagon.

In addition to their studies on the effect of the changes related to His1 and Asp9 of glucagon on the binding and activity, Unson and Merrifield (26) investigated the effect of the replacement of some serine residues. Indeed, as mentioned before, the His, Asp, Ser triad can make up an active center, similar to the one in serine protease. Therefore, they tested the requirement for serine by a series of substitutions of the Ser2, Ser8, Ser11, and Ser16 residues. They synthesized 38 peptides containing replacements at these positions and tested them for binding affinity and for adenylate cyclase activity. Replacement of Ser2 by Tyr or Phe resulted in almost total loss of binding and activity. However, when Ser2 was replaced by Ala or Thr, the binding remained strong, and so did the activity. When these replacements were coupled with the deletion of His1, there was practically no binding and no activity. The substitution of Ser8 by Ala, Gly, Thr, and Asn reduced greatly both binding and activity. Again, when coupled with the deletion of His1, binding and activity became very small.

When Ser11 was replaced by such residues as Ala, Thr, and Asn, there was no decrease of the binding, and the analogues so obtained still elicited a high cAMP response. As a matter of fact, replacing Ser11 by Ala or Thr produces analogues that bind to the receptor respectively 4 and 2.5 times better than glucagon.

Serine modifications at the 16 position produces analogues that retained their binding ability, with the exception of Thr16 glucagon amide. Indeed, it becomes apparent that a bulky side chain at that site is unfavorable to the binding. As far as the adenylate cyclase activity is concerned, the Ser16 substitutions produce analogues lacking activity. This holds true even when L-Ser16 is replaced by D-Ser. This is another proof that the requirements for binding to the receptor are less rigorous than the requirements for the transduction signal. The Ser16 residue thus appears to be necessary for transduction, probably related to the position of the OH group at a very specific site.

The presence of a crucial serine residue, probably part of the His, Asp, Ser triad, raised the question of whether an activated serine is present in the glucagon-receptor complex. To attempt to answer this question, Unson and Merrifield (26), treated glucagon in the presence of hepatocyte membrane with di-isopropylfluorophosphate (DIPF) or with 4-amidinophenyl-methane sulphonyl fluoride (APMSF). The DIPF reduced the activity to 22%, while the APMSF reduced it to less than 0.1%. These chemicals, which are serine protease inhibitors, also proved to be inhibitors of glucagon. Glucagon alone, without the receptor, and a membrane preparation containing the glucagon receptor, the G protein, and adenylate cyclase (but no glucagon) were not affected by the inhibitors. It was thus found (26) that the decrease of adenylate cyclase activity by the serine protease inhibitors requires a mixture of glucagon and its receptor system.

It was thought that the entire sequence of glucagon is required for activity. However, the 1 to 6 residues fragment of glucagon demonstrated activity by itself (27), leading to the conclusion that the N-terminus half of the molecule is responsible for transduction and the carboxyl half for binding. In agreement with this hypothesis was the important role played by His1 and Asp9 in transduction, and their interaction with a site in the receptor leads to activation.

The investigation of the role of serine enforces the similarity with the serine protease His, Asp, Ser triad. Unson and Merrifield (26) propose that this triad functions via the formation of an oxyanion formed from the hydroxyl group of serine through the abstraction of the proton by a histidine residue in a charge relay network. Consequently, the oxyanion performs a nucleophilic attack on the substrate.

The results of the study (26) show serines 2, 8, and 11 to be more involved in binding than in activity since the activity decrease upon replacement of one of these residues is due to the decrease in binding. In addition, the Ser2 and Ser11 residues do not seem to be crucial for binding. The Ser16, in contrast, does not seem to be strongly involved in the binding, but it turns out to be the third important residue for transduction. Since Ser16 is located in the middle of the hormone, in one of the predicted beta turns, it was found (26) that His1, Asp9, and Ser16 can be juxtaposed to create the charge relay for the triad. The presence of a receptor is necessary for glucagon to be inhibited by serine protease inhibitors. It is possible, thus, that glucagon in the presence of the inhibitor assumes the conformation in which Ser16 is activated (28). The glucagon-receptor complex can acquire enzymatic activity by aligning the His, Asp, Ser triad for amine bond hydrolysis, and it would position a sensitive bond in the receptor for enzymatic cleavage. New carboxyls and amino groups can be liberated and activate the system for reaction with the G protein, thus starting the cascade of events induced by glucagon.

A structural study of the glucagon receptor (28) has reported that a 64 kDa and a 33 kDa fragment were labeled with radioactive iodoglucagon.

Indeed, the glucagon receptor is a 63 kDA transmembrane glycoprotein that can be cleaved by proteolytic enzymes to give 33 kDa and 24 kDa segments that still can bind glucagon (28). When bound to glucagon, the 33 kDa segment of the complex was found to be sensitive to GTP. An agonist bound to the receptor may form an active center, and an antagonist may not bring His, Asp, Ser in the proper conformation of might miss a required residue.

As shown before, it was proven that Asp9 is an essential residue for activity. Unson et al. (29) have also investigated the role of Asp15 and Asp21 residues in glucagon.

The presence of the His1 residue is also necessary for transduction, and it has been proposed, as shown before, that histidine interacts with Asp9. However, since the glucagon molecule is quite flexible, this interaction might take place between histidine and Asp15 or Asp21. It was found (29) that the substitution of the Asp15 residue decreases the binding of glucagon to the receptor. The replacement of Asp15 by a neutral residue, such as leucine or norleucine, reduces the binding by about 90%. If a positive residue such as Lys is introduced at the 15 position, the binding is completely lost. A particularly interesting result is that if L-Asp15 is replaced by D-Asp15, the binding is significantly decreased but the activity remains as strong as in normal glucagon.

Replacing Asp15 by another negative residue, such as Glu, reduces somewhat both the binding and the activity but only to about 80% (as far as the activity is concerned, this is true only at very high concentrations). In general, with few exceptions, the loss of activity follows the loss of binding.

As far as Asp21 is concerned, the results are different. Many of the replacements lead to analogues that retain their binding affinities. The analogues featuring Leu21, Gln21, and Glu21, and an amide group instead of the terminal carboxyl bound 236%, 29.5%, and 89.1% respectively. They were as potent as natural glucagon. When des His1 analogues were prepared, they bound well but showed marked decease in activity. Thus, it looks as though position 21 plays a role in transduction but less so in binding. Analogues with replacements at both the 9 and 21 positions showed weak activity and no activity when the His1 residue was also deleted. Those analogues, which did feature Asp9 but no Asp21, still showed activity, indicating that one of these Asp residues is necessary, preferably Asp9.

These data (29) suggest that glucagon is flexible enough to adopt a number of conformations that can bind to the receptor. Not all these conformations will show transduction activity, but the peptide is small enough to allow many residues to occupy sites in proximity to each other, with distances sufficient for weak interactions.

Asp9 thus remains the crucial aspartic residue for transduction, and it is a point of detachment of the receptor binding function from adenylate cyclase activity. The Asp21 residue is not essential for transduction or binding, while more than 96% of the cAMP formation disappears, with the same decrease in receptor binding if the Asp15 residue is replaced. It can be inferred that the histidine residue can interact with an Asp residue either at site 9, 15, or 21, even though 9 is preferred, with 21 the weakest. However, the contribution of the 15 position to the adenylate cyclase activity is too related to the binding for a clear picture of the role of this residue to emerge (29).

Additional studies of the glucagon receptor were performed by Unson et al. (30) by preparing polyclonal antibodies against synthetic peptides corresponding to four different extramembrane segments of the rat glucagon receptor.

The PR-15 and DK-12 antibodies were directed against a synthetic peptide corresponding to amino acid residues 103–117 and 126–127, respectively, of the N-terminal tail. The KD-14 antibody was directed against the synthetic peptide corresponding to the residues 206–219 of the first extracellular loop, and the ST-18 antibody was directed against the residues 468–485 of the intracellular C-terminus. The DK-12 and KD-14 blocked the binding of glucagon to the receptor but did not increase the activity of adenyl cyclase stimulation. On the contrary, they showed some antagonistic activity. The PR-15 and the ST-18 antibodies showed no effect on either binding or activity, indicating that the respective fragments of the receptor are not involved in either.

Ab initio calculations are being performed at present in order to study the histidine–aspartic–serine triad.

References

1. Stryer, L. *Biochemistry*, 4th ed. W.H. Freeman and Co., New York, 1995.
2. Unson, C.G., Gurzenda, E.M., Iwasa, K., and Merrifield, R.B. *Journal of Biol. Chem. 264*, 2, 789, 1989.
3. Sasaki, K., Dockerill, S., Adamiak, D.A., Tickle, I.J., and Blundell, T. *Nature 257*, 751, 1975.
4. Chou, P.Y., and Fassman, G.D. *Adv. Enzymol. 2*, 45, 1978.
5. Chou, P.Y., and Fassman, G.D. *Biochemistry 14*, 2536, 1975.
6. Unson, C.G., Andreu, D., Gurzenda, E.M., and Merrifield, R.B. *Proc. Natl. Acad. Sci. USA 84*, 4083, 1987.
7. Andreu, D., and Merrifield, R.B. In *Peptides: Structure and Function*, Deber, C.M., Hruby, V.J., and Kopple, K.D. (eds.) Pierce Chem. Co., Rockford, IL, 595, 1985.
8. Andreu, D., and Merrifield, R.B. *Eur. J. Biochem. 164*, 585, 1987.
9. Wakelam, M.J.O, Murphy, G.S., Hruby, V.J., and Houslay, M.D. *Nature 323*, 68, 1986.
10. Trimble, E.R., Bruzzone, R., Bidin, T.J., Meehan, C.J., Andreu, D., and Merrifield, R.B. *Proc. Natl. Acad. Sci. USA 84*, 3146, 1987.
11. Stewart, J.M., York, E.J., Baldwin, R.L., and Shoemaker, K.R. 19th European Peptide Symposium 1986, Porto Carras, Greece (abs.).
12. Lu, G.S., Mojsov, S., and Merrifield, R.B. *Int. J. Pept. Prot. Res. 29*, 545, 1987.
13. Korn, A.P., and Ottensmayer, F.P. *J. Theor. Biol. 105*, 403, 1983.

14. Chung, F.Z., Wang, C.D., Potter, P.C., Venter, J.C., and Frase, C.M. *J. Biol. Chem.* *263*, 4052, 1988.
15. Gronenborn, A.M., Boverman, G., and Core, G.M. *FEBS Lett.* *215*, 88, 1988.
16. Unson, C.G., Macdonald, D., and Merriefield, R.B. *Arch. Biochem. Biophys.* *300*, 747, 1993.
17. Merriefield, R.B., and Unson, C.G. *Peptides Proc. Chinese Peptide Symp., Hangzhou, China 251*, 1993.
18. Unson, C.G., and Merrifield, R.B. *Proc. Natl. Acad. Sci. USA 92*, 454, 1994.
19. Schrag, J.D., Li, Y., Wu, S., and Cygler, M. *Nature 351*, 761, 1991.
20. Post, S.R., Rubinstein, P.G., and Tager, H.S. *Proc. Natl. Acad. Sci. USA 90*, 1662, 1993.
21. Berstein, F.C., Koetzle, T.F., Williams, G.J.B., Meyer, E.F.Jr., Brice, M.D., Rogers, J.R., Kennard, O., Shimanouchi, T., and Tasumi, M. The protein data bank. *J. Mol. Biol. 112*, 535, 1977.
22. Cramer, C.J., and Truhlar, D.J. *AMSOL Program 606, QCPE.* Indiana Univ. Bloomington, Indiana, 1992.
23. Carruthers, C.J.L., Unson, C.G., Kim, H.N., and Sakmar, T.P. *J. Biol. Chem. 269*, 18, 29321, 1994.
24. Jelinek, L.J., Lok, S., Rosenberge, G.B., Smith, R.A., Grant, F.J., Biggs, S., Bensch, P.A., Kuijper, J.L., Sheppard, P.O., Sprecher, C.A., O'Hara, P.J., Foster, D., Walker, K.M., Chen, L.H.J., McKernan, P.A., and Kindsvogel, W. *Science 259*, 1614, 1993.
25. Iwanji, V., and Vincent, A.C. *J. Biol. Chem. 265*, 21302, 1990.
26. Unson, C.G., and Merrifield, R.B. *Proc. Natl. Acad. Sci. USA, 91*, 454, 1994.
27. Wright, D.E., and Rodbell, M. *J. Biol. Chem. 254*, 268, 1979.
28. Iyengar, R., and Herberg, J.T. *J. Biol. Chem. 259*, 5222, 1983.
29. Unson, C.G., Wu, C.R., and Merrifield, R.B. *Biochemistry 33*, 6884, 1994.
30. Unson, C.G., Cypress, A.M., Wu, C.R., Goldsmith, P.K., Merrifield, R.B., and Sakman, T.P. *Proc. Natl. Acad. Sci. USA V93*, P 310, 1996.

9
The Alpha Factor

The alpha factor is a 13 amino acid peptide secreted by some cells of yeast. The yeast, *Saccharomyces cerevisiae*, is a eukaryote that exists either in a haploid or diploid state. It can undergo both sexual and asexual reproduction. The former occurs via the mediation of the mating of two haploid cells, types *a* and alpha, by two diffusible peptide pheromones, *a* and alpha factors. Each haploid responds to the appropriate pheromone by a series of responses such as cell-surface agglutinin synthesis, cell-cycle arrest in G1, and morphological transformation necessary for cell and nuclear fusion (1). The two types of haploids can be considered two genders since they mate. After the fusion of an *a* cell with an alpha cell, a zygote is formed that can propagate into a diploid line by budding or can undergo meiosis and sporulation, giving rise to a new haploid phase of the cycle (2–3).

The recognition of the two cells takes place via a biochemical process: The *a* cells release S-Farnesylated and carboxy-terminus methylated dodecapeptide NH2-Tyr-Ile-Ile-Lys-Gly-Val-Phe-Trp-Asp-Pro-Ala-Cys (Farnesyl)-COOMe, which is called the *a* factor.

The alpha cells produce the tridecapeptide NH2-Trp-His-Trp-Leu-Gln-Leu-Lys-Pro-Gly-Gln-Pro-Met-Tyr-COOH, which is called the alpha factor.

The alpha factor has no effect on the alpha cells, but it has significant effect on the *a* cells. As mentioned above, this effect consists in stopping the cell's growth in the G-1 phase of the cell's cycle, via a mechanism that inhibits DNA replication, increases cell surface agglutinability toward the cells of the other mating type and produces an aberrant elongation of the *a* haploids. When alpha cells are in contact with the *a* factor, similar effects are produced (4).

In order to produce these effects, the alpha factor has to bind to a receptor that is located in the *a* cell membrane. The receptor is a 431 residues protein, coded by a gene called STE-2 (5–6), with 7 folds made out of four extracellualr domains, four intracellular domains, and seven membrane domains. The signal transduction proceeds via an interaction between the factor–receptor complex and the G-protein, similar to the transduction signal of glucagon (see Chapter 8), rhodopsin (7), and the nicotinic acetylcholine receptors (8). In the alpha-factor case, the conformational charge of the pheromone leads to an activation of the G-protein, resulting in a dissociation of the G alpha subunit from the G beta subunit.

A large number of studies have been performed to determine the alpha-factor conformation (9). Among these, an NMR study performed by Jelicks et al. (10) has shown that the solution structure of the alpha factor contains a type 11 beta turn involving residues 7–10, a turn that does not appear in the inactive peptides studied. No other long-lived structured conformation was found in that study, in disagreement with the results of Higashijima et al. (11), who found for the peptide a defined structure in solution, comprising the beta turn and a helical N-terminus.

Jelicks et al. (12) attempted to elucidate the structure of the alpha factor as part of the peptide-receptor complex, and as such investigated the influence of phospholipids on the conformation of the pheromone. To this purpose they used NMR, especially NOESY methods, to obtain information on the structure of the alpha factor bound to phospholipid vesicles. In addition, they studied the interaction of the peptide with the head group of the lipids. One group of the lipids investigated was neutral, and another group was anionic. These groups of lipids are present in the plasma membrane of S. cerevisiae (12). The temperatures at which the alpha factor was added to the vesicles were above the lipid phase transition temperature, T_c. The systems were cooled when the interaction with the gel phase was measured. The membrane system was modeled by sonicated phospholipid vesicles.

Quasielastic light scattering (QLS) measurements were performed on the alpha-factor–vesicle mixture and on the vesicles alone. It was found that upon addition of the alpha factor, the vesicles increased in size. Their polydispersity increased for saturated lipids and decreased for unsaturated ones (12). In NMR experiments, addition of lipids to the alpha factor led to a broadening and upfield shifting of most peptide proton resonance.

Apparent affinity constants K_a were calculated using the expressions

$$K_A = M_F / [(1 - M_F)(L - M_F P_0)],$$

$$M = (\delta_{OBS} - \delta_{FREE}) / (\delta_{BOUND} - \delta_{FREE}),$$

where M_f is the mole fraction of the bound peptide, L is the molar concentration of the lipid, P_0 is the initial molar concentration of the peptide, δ_{free} is the chemical shift of the free peptide, and δ_{bound} is the chemical shift of the bound peptide. The peptide was assumed to be fully bound when there was no more change in the chemical shift upon addition of more alpha factor.

The NOESY spectrum was used to investigate the conformation of the alpha factor bound to vesicles.

Experiments with PrCl3 and L-alpha-dipalmitoylphosphatidylcholine (DPPC) vesicles show that the alpha factor interacts with the outer leaflet of the phospholipid vesicle but not with the inner leaflet phospholipid (13).

QLS and P experiments show that the alpha factor interacts with the lipid vesicles. The NMR studies show that the apparent affinity of the pheromone for the negatively charged vesicles is much greater than that for neutral lipids (12).

The NOESY studies indicate that the peptide is quite flexible and confirm the fact that the central portion of the molecule forms a type II beta turn, which seems to be essential for biological activity (12).

The presence of the beta turn is particularly confirmed by the NOE observed in solution at the Pro8-Alpha-CH-Gly9 NH site. This NOE, as well as the one between the Gly9 NH and the Gly9 alpha CH, are very intense when the peptide is bound to the lipid. This demonstrates that the beta turn is more stable in the peptide–lipid complex (12).

The conformation of the system as determined by Jelicks et al. (12) resembles and differs in certain points from the one proposed by Wakamatsu (14). The latter propose a possible 3 helix at the N-terminus that inserts in the bilayer and an extended structure for the 6–9 residues. The amphiphilic behavior for the N-terminus proposed by Wakamatsu et al. is consistent with the NOE results. However, the results of Jelicks et al. (12) leave no doubt about the presence of a beta turn at the 7–10 residues. The discrepancies between the two sets of results could be attributed to the fact that Wakamatsu et al.'s (14) experiments were performed only in D_2O, while the others were performed in both D_2O and in H_2O, involving thus the exchangeable amide protons.

In order to test further the hypothesis that the alpha factor is a flexible peptide that assumes a folded conformation as the conformation featuring biological activity, Xue et al. (15) synthesized alpha-factor molecules in which constraints were placed on the distributions of permitted conformations. A cyclic analogue was synthesized in which an amide bond was formed between the side chains of residues 7 and 10. This bond between side chains causes a bend for the Lys7-Pro8-Gly9-Gln10 sequence, which is believed to form a beta turn of type II. It is thus of interest to study the biological activity of this cyclic peptide.

Analysis of the synthesized analogue indicated that the molecule is indeed a tridecapeptide with all the residues from alpha factor present (15). Mass spectroscopy experiments confirm these findings and indicate that the peptide is cyclic. The molecule was found to be highly active in arresting the growth of cells in a wild-type tester, *S. cerevisiae* 2180-1A. In rapport with different testers, the cyclic peptide showed activities of one-fourth, one-fifth, or one-twentieth of the linear peptide. Since the cyclic peptide features an acylated lysine side chain, it was of interest to study the activity of the linear peptide featuring an acylated lysine residue. This analogue exhibited an intermediate activity between the cyclic peptide and the natural linear peptide (15).

The recovery of the strain 2180-1A was examined by determining the proportion of growth-arrested cells in cultures exposed to the cyclic peptide and to the linear peptide. It was found that the start of recovery was 12 hours for the cyclic peptide and 14 hours for the linear one. HPLC analysis has demonstrated that there is no evidence of the breakage of the cyclic peptide in the first 4 hours (15). The experiments show no evidence that the cyclic peptide reverts to a linear form. It can thus be understood that the activity is due to the cyclic peptide itself and not to an eventual linear molecule formed by the hydrolysis of the amide bond (15).

It is known that des-Trp1-Ala3, Nle12 alpha factor is an antagonist of the alpha factor (2–3), which binds to the receptor but does not feature biological activity. Xue et al. (15) tested the antagonistic strength of this analogue against the cyclic peptide. Indeed, they found that the antagonist inhibits the biological activity of both Nle12 alpha factor and of cyclo Nle12 alpha factor, indicating thus that the cyclic peptide binds to the same receptor as the native pheromone.

It is somewhat difficult to understand why a molecule featuring a bend at the site where a beta turn of type II seems to be necessary for activity shows a smaller activity than the native pheromone. However, it is clear that the cyclic peptide binds to the receptor and features biological activity. It is possible that the drop in activity is due to the fact that the linkage between Lys7 and Glu10 might prevent optimal binding to the receptor (15), especially since interactions between the termini and the receptor are crucial.

It has been shown that the amino acids in the primary structure of the peptide have a great influence on activity (16). For example, His2 and Tyr13 are essential for activity (16). But Xue et al. (15) proved that the Lys7 (ac), Nle12 alpha factor is almost as active as the native alpha factor, thus demonstrating that a chargeable amine at position 7 is not essential for activity.

As shown by Xue et al. (15), antagonists are not satisfactory tools for the investigation of the conformation of an active hormone at the receptor. Indeed, they bind to the receptor but do not elicit biological response. In consequence, an agonist that binds to the receptor and shows biological activity provides better information not only about the requirements for binding but also for those for exhibiting activity.

The studies of Matsui et al. (17) and Naider and Becker (2) have shown that His2 in the alpha factor is essential for activity. Indeed, deletion of this residue decreased the activity of the hormone 10 times, and replacement by Leu, Phe, and D-His inactivated it completely. Levin et al. (18) hypothesize that the protonation of the His2 residue may stabilize the conformation of the pheromone that ensures maximum binding. A truncated pheromone (desTrp1, desHis2, Nleu12) competed with the native pheromone for the binding, but did not show any activity, proving thus to be an antagonist (19).

Histidine often plays an important biological role, as for example in the activity of glucagon (see Chapter 8). Indeed, its imidazole ring can act as an acid or a base, facilitating proton transfer in some biological reactions. Histidyl residues may enter into electrostatic networks that stabilize the biologically active conformation of the peptide–receptor complex, playing a role in the triggering of the signal.

Levin et al. (18) carried out a detailed study of the importance of histidine for the pheromone activity and binding. This study consisted of synthesizing some analogues of the alpha factor, all them containing norleucine instead of methionine at the 12 position. This was done for purposes of facilitating the synthesis, since the native alpha factor and the Nleu12 pheromone exhibit the same activities (18). The analogues examined contained beta thienylalanine, 1-methylhistidine, 3-methylhistidine, or 3-pyridylalanine instead of histidine at the 2 position. The 1-methyl and 3-methylhistidine residues were investigated in both L and D chirality. The activity of the analogues was tested as cell-growth stopping ability against three types of cells: 2180, RC629, and 50bts (18).

It was found (18) that none of the D-containing analogues showed activity against any of the strains. The authors concluded that all D analogues are antagonists. The methylation of histidine resulted in a 5- to 10-fold drop in activity, while the replacement of the imidazole ring from histidine by a thiphene ring resulted in a 5- to 100-fold drop in activity. When the imidazole ring was replaced by a pyridil moiety, only a 2- to 5-fold decrease in activity was observed.

Even though histidine residues have been considered essential for the biological activity of many systems and have shown the ability to influence the binding to protein receptors (18), from the results of Levin et al. (18) it can be seen that the imidazole side chain of His2 is not crucial for the biological activity of the alpha factor. Indeed, even though the activity is greatly reduced for some replacements, in other cases it is simply halved, such as in the case of the replacement by a pyridil ring. The replacements studied, with the exception of the thiophene replacing histidine, are of the type that does not preclude protonation. However, the results show that protonation is not essential for binding.

The antagonism of the D analogues is attributed to the fact that they displace the native pheromone from the receptor but do not trigger the signal for biological activity. As in the studies on glucagon (see Chapter 8), the binding and biological activity of hormones can be uncoupled. In the case of the alpha factor, while the histidine residue at position 2 does not seem to play an important role in the binding, its ability to trigger the mating cascade is dependent on both the atoms present in the ring and on the chirality (18). When chirality is changed from L to D, the binding is not affected but the activity is decreased. If the two nitrogens of histidine are replaced by carbon or/and sulfur, the activity is significantly decreased. However, when the imidazole ring is replaced by a pyridyl ring, the activity decreases only 2- to 5-fold (18). It appears thus that the nitrogen atom plays a particularly important role in the triggering of the signal.

Some studies (10,12) of the des-Trp1 dodecapeptide alpha factor analogues have shown that those that feature a type II beta turn exhibit higher activity than those analogues containing residues that prevent the formation of the turn. Among these, the D-ala9 alpha factor was found via NMR experiments to feature a transient type II beta turn, containing the 8 and 9 residues, while linear analogues containing an L residue in the 9 position did not show a defined structure in solution (12). Other studies have found the existence of turns in the center of the pheromone and at the carbon terminus, while the amino terminus site was alpha helical (20). Garcia-Echeveria et al. (21) have shown that a cyclo Cys, Pro, D-Val, Cys tetrapeptide features a type II beta turn in water.

In order to investigate further the importance of beta turns for the biological activity of the alpha factor, Gounarides et al. (22) have synthesized a number of analogues of the pheromone in which the 7–10 residues are replaced by cyclo Cys, Pro, X, Cys units. The cycle is constrained by a S–S bond. It is found that these peptides have a higher activity than their linear homologues (22).

These systems were investigated by Gounarides et al. (22) with NMR experiments and with molecular modeling methods. The cyclo Cys, Pro, X, Cys tetrapeptide fragments (where X is Gly, L-Ala, D-Ala or D-Val) were geometry optimized using the AMBER force field (23) and the SYBYL software package. The energy minimizations were performed with the Newton–Raphson and gradient methods (24) to a threshold of 0.001 kcal/mole. Since the cyclic peptides so examined are significantly more constrained than their linear counterparts, the interpretation of the NMR data is simpler. It was concluded that the cyclic disulfide regions of the analogues feature turn structures both in water and in DMSO/water mixtures.

According to the NOE results, the cyclo Cys, D-Val, Cis, Nle alpha factor and the Cys, D-ala, Cys, Nle species were thought to form a type II beta turn. However, the L-Ala containing species was thought to exhibit a type I beta turn. The Gly9 analogue favored a type II turn in the DMSO/water mixture (22).

The activity of these analogues follows a trend where it is apparent that the presence of a turn may be more important for the biological activity of the alpha factor analogue than the conformation it assumes.

In order to investigate further the conformation of the alpha factor, both free in solution and bound to phospholipid vesicles, Garbow et al. (25) have applied high-resolution proton NMR analysis.

They applied the ^{13}C, ^{15}N rotational-echo double resonance (REDOR) NMR method, which is a modern solid-state NMR experiment. This method makes it possible to obtain accurate measurements of the carbon–nitrogen distance in solid samples (26–27). This method has been used to study the tripeptide melanostatin (28) and gramicidin A in multilamellar dispersions (29).

Garbow et al. (25) synthesized four alpha-factor analogues labeled with ^{13}C or ^{15}N. The labeling was introduced at positions based on an examination of the folded and extended conformations of the tridecapeptide by computer modeling.

To help interpret the REDOR data, Garbow et al. (25) performed a geometry optimization of the Nle12 alpha factor analogue, which is a very flexible molecule, using the CHARMm22 program. For a flexible molecule, molecular modeling produces a high number of local minima, and in order to assess the global minimum, a number of initial geometries have to be tested, especially where the dihedral angles are concerned. Since previous NMR studies have suggested a beta turn at the Pro8-Gly9 residues, and since the value of the Pro8 phi angle is restricted by the proline ring, the dihedral angles of Gly9 are most important in the description of a possible turn in this region of the pheromone.

Three starting conformations were used:

- With a type II beta turn centered at Pro8 and Gly9;
- With a type I beta turn centered at Pro8 and Gly9;
- A fully extended conformation.

Prime-type turns were not considered, since the presence of an L-proline residue precludes their existence. The energy minimization was performed in the following way: Three ideal conformations were used as starting geometries and optimized. The interatomic distances so obtained were subsequently compared to the REDOR results. Garbow et al. (25) found that the extended conformation does not contribute significantly to the structure of the alpha factor in the lyophilate. Indeed, the interatomic distances corresponding to the extended structure differ by about 1 A from the REDOR results. The closest values to the REDOR values correspond to the conformation featuring a type I beta turn. However, the type II beta turn optimized conformation also shows interatomic distances close to the REDOR ones (within 0.4 A) (25).

Garbow et al. (25) also compared the REDOR-obtained interatomic distances with the distances calculated for a random distribution of structures in the Lys7-Gln10 region, excluding all dihedral angles that violated Van der Waals radii. This search leads the authors to conclude that the beta turns play a major role in the distribution of the alpha factor in solid powder. The most probable appears to be a distorted type I beta turn.

In general, the REDOR results from the study indicate that the Nle12 alpha factor shows a preference for bent structures even in liophilized powder form. This result confirms the previous NMR results that the alpha factor assumes a type II beta turn conformation in organic and water media as well as bound to lipid vesicles. It is true that the lyophilate of the enzyme retains a quantity of water. However, there will exist more interpeptide interactions than in solution. Still, the preference for turns indicates that the peptide has a strong tendency to bend (25). It is also clear that the REDOR method provides an important tool for the study of peptides in solid state.

Marepalli et al. (30) used molecular modeling to examine the influence of side-chain lactamization on the freedom of the backbone residues. Indeed, cyclization of peptides reduces conformational freedom, and it might contribute to arranging the peptide in a conformation favorable to receptor binding. As a consequence of cyclization, the resulting conformational homogeneity can lead to an increased receptor specificity, to an increase of the agonistic or antagonistic potency, or to increased resistance to degradation (31–33). The type of cyclization investigated for the peptides has included chain end to chain end, side chain to end, and side chain to side chain. The former can inactivate the molecule due to the fact that the termini cannot bind to the receptor (34). However, the last two types of cyclization have resulted in many cases in quite potent analogues (35). Examples of such cyclizations include the obtaining of potent and selective opioid peptides via side-chain-to-side-chain i to $i + 2$ amide bond formation (36), cyclization of the i to $i + 4$ side chains of the growth hormone releasing factor analogues (37) and a human calcitonin analogue (38) and others; i to $i + 3$ lactamization gave high activity and binding to cholecistokinin analogues (39) and increased anticoagulant activity to a thrombin-binding hirudin analogue (40). However, conclusive evidence about the relation between the size of the ring and activity was not obtained in the above-mentioned studies.

Yang et al. (41) have reported that the biological activity and the receptor-binding ability of ciclo alpha factor analogues in which there are amide bonds between residues in the 7–10 region depend on the size and composition of the rings. Marepalli et al. (42) have synthesized N-acetyl carboxyl amide terminal cyclic tetrapeptides with the X-Pro-Gly-Z sequences, where X = Lys, Orn, Dab (alpha, gamma diaminobutyric acid) or Dpr (alpha, beta diaminopropionic acid) and Z = Glu or Asp. CD studies have indicated that these peptides feature beta turns centered probably on the Pro-Gly dipeptide. To obtain more information about the structure of these peptides, Marepalli et al. (30) performed two-dimensional studies and molecular modeling.

The molecular modeling methods included the BIOPOLYMER module of the SYBYL software and the AMBER force field used for minimization. The minimization and the constrained searches were performed using the annealing procedure in SYBYL (30). The DMSO molecule was optimized using the ab initio method with the 6-31G* basis set. The charges and geometrical parameters so obtained were introduced into the Kollman All-Atom parameter set of SYBYL (30). To examine the solvent environment of the tetrapeptides, boxes of DMSO molecules were built. These boxes were almost cubic (30), with lengths between 31 and 36 A.

Molecular-dynamic simulations in vacuum and DMSO were performed on the tetrapeptides using the Amber field and the Verlet program (30) at constant temperature.

It would be expected for these peptides to feature a beta turn of type II, but the situation, as reported by Marepalli et al. (30), is more complex. For type I and type II beta turns, an interprotonic distance of 2.4 A between the NH groups of the i + 2 and i + 3 residues is expected. Types I and II beta turns differ by the value of their torsion angles psi2 and phi3. In the former, these angle are –30° and –90°, while in the latter they are 120° and 80°, respectively. A gamma turn (see Chapter 10) has also to be considered. The cyclo (Ac-Lys-Pro-Gly-Glu-NH$_2$) peptide, which features a 1,4 bond, seems to feature a gamma′ turn. When instead of Lys there is an Orn residue, a gamma turn conformation is observed. However, if Lys is replaced by Dab, the molecule features an almost perfect type II beta turn. The same structure is found for the peptide in which Lys has been replaced by Dpr. However, this molecule exhibits oscillations between this structure and two structures featuring gamma and gamma′ turns. The lysine-containing analogue with Glu replaced by Asp features a type II beta turn. The peptide containing both Dab and Asp features extensive hydrogen bonding. It adopts more than one conformation, but the most stable seems to be the one featuring a gamma ring. The molecule with Drp and Asp is the most constrained of the series, and as such, experiences less interaction with the solvent. The NMR-obtained interproton distances indicate the presence of the type II beta turn.

Previous studies have indicated that an i to i + 3 lactam bridge is helix destabilizing (43). In contrast, concerning a beta turn, the results of Marepalli et al. (30) suggest that the exact conformation depends more on the composition of the rings than on their size. If these models are to be used as conformation building blocks for larger peptides that might be alpha-factor analogues, the choice of the specific residues involved is a crucial element in obtaining a beta turn. Since the presence of a beta turn is important for biological activity, both the size of the ring and the nature of the residues have to be considered. It was also concluded that none of the peptides examined features a type I beta turn (30).

Yang et al. (41) showed that the constrained tridecapeptides retain high biological activity even though they do not bind to Ste2p with high affinity. Antohi et al. (44) attempted to understand better the relation between high biological activity of the alpha factor analogues, their receptor affinity, and their conformation. In order to achieve this goal, they studied four constrained tridecapeptides that feature an i to i + 3 lactam in the residues X-Pro-Gly-Glu, in which X can be lysine, ornithine, Dab, or Dpr. This study makes use of NMR and molecular modeling methods.

The NMR studies include COSY, ROESY (rotating frame nuclear Overhauser effect spectroscopy), and NOESY (nuclear Overhauser effect spectroscopy).

The molecular modeling consists in building the analogues with the BIOPOLYMER module of SYBYL, creating templates for the residues used in cyclization with the dictionary options of the SYBYL software (44). The searches created 200 initial conformations for each molecule, with constraints derived from the NOESY spectra. The structures were minimized using the AMBER force field, and their deviations from the physical chemical data were estimated by computing rms deviation values of the constrained distances (44). Ten conformations of each molecule were selected for further investigation and were subjected to energy minimization using annealing and minimization routines (44). The lowest-energy conformation thus obtained was encapsulated in a solvent box, and the energy was minimized in the presence of the solvent, which was DMSO. The box used was the smallest box in which periodic boundary conditions could be respected. This box was a cube with sides between 35 and 42 A (44). Dihedral angles obtained from NMR coupling constants were not used as constraints but were used for evaluating the model obtained from the NMR distance constraints.

The four peptides investigated by Antohi et al. (44) feature the following sequence:

Trp-His-Trp-Leu-Gln-Leu-NH-CH-CO-Pro-Gly-NH-CH-CO-Pro-Nle-Tyr

According to the values of m and n, they are called C_{42}, C_{32}, C_{22}, and C_{12}, where C_{42} corresponds to $m = 4, n = 2$, C_{32} corresponds to $m = 3, n = 2$, C_{22} corresponds to $m = n = 2$, and C_{12} corresponds to $m = 1, n = 2$.

The absence of a NOESY peak between $i + 2$ and $i + 3$ suggests that the $i + 3$ residue amide proton points outside the beta turn. The values of psi and phi of the $i + 2$ residue (Gly) confirm this conclusion. A seven-member ring is observed in the molecular-dynamics calculations. It was found that the major conformation of C_{42} is a gamma' ring. The Gly9 residue is rather extended.

The C_{32} analogue exhibits NOE results, which seem to indicate the absence of a type II beta turn. The dynamics simulations confirm this result and reveal the presence of a seven-membered ring. The average torsional angles in this analogue correspond to a gamma turn.

In the C_{22} analogue the interatomic distances found with the NOE experiments suggest the presence of a type II beta turn. In addition, the molecular-dynamical simulations seem to indicate that the molecule can adopt a conformation with a transient gamma turn centered at the glycine residue.

In C_{12}, as in the model tetrapeptide, three types of turn, gamma, gamma', and type II beta, can coexist.

When attempts were made to correlate the above findings with the biological activity of these analogues by measuring the growth arrest activity

(44), it was found that all four exhibit the same activity, that is, 10 times less than the linear tridecapeptide. This result indicates (44) that neither of these structures is present in the native alpha factor. Moreover, these analogues have much lower receptor activity than the linear alpha factor. However, the fact that they do possess a certain amount of activity may suggest that the alpha factor does feature a bent conformation in order to bind to the receptor and for the transduction of the signal.

In order to investigate the role of the Gly9 residue, Shenbaghamurti et al. (45) and Naider et al. (46) have replaced it in the alpha factor by D-Ala and L-Ala. The analogue featuring D-Ala instead of Gly9 was equally active as the native alpha factor. In the case of the replacement of Gly9 by L-Ala, the activity was reduced tenfold. The difference in activity was correlated with the relative tendencies of the D-Ala and L-Ala analogues to form beta turns, as determined by NMR studies (47) and vibrational circular dichroism (48). Indeed, while the D-Ala containing alpha factor analogue featured a transient type II beta turn as observed by both NMR and vibrational circular dichroism experiments, the L-Ala containing tridecapeptide presented a short-lived type I beta turn conformation, which could be observed only by the short-time-scale technique of vibrational circular dichroism (48).

The secondary structure of a peptide is the result of local steric and electronic effects, hydrogen bonding, and interactions with the medium in which the peptide chain is solvated (49). Concerning the alpha factor, Antohi et al. (50) suggest that it is the local effect of the methyl side chain that is responsible for the significant difference in activity between the D-Ala9 and the L-Ala9 alpha-factor analogues.

In order to obtain further information on the structure of these two optical isomers, Antohi et al. (50) applied ab initio quantum-chemical methods to the study of the effect of changing the chirality of an alanyl residue in the 9 position of the alpha factor.

The conformations investigated are the L-Pro-Gly (as in the native alpha factor), L-Pro-D-Ala, and L-Pro-L-Ala. These dipeptides are not blocked, and in consequence, they cannot span a complete turn region. However, by investigating the order of stability of different conformations of these dipeptides, one can gain more information about the energetic basis of the behavior of the alpha-factor analogues. Indeed, the structural differences between the central regions of the analogues previously examined (47–48) is the same as the difference between the dipeptides.

Previous chapters have discussed the structures of proline, glycine, and alanine. As far as the dipeptides are concerned, there has been an attempt (50) to find the geometry optimized conformations associated with all the local minima. In addition, the energies of the three dipeptides were calculated for various conformations.

The Gaussian-90 computer program was used to perform ab initio calculations with the 6-31G basis set. The geometry optimization was performed using the gradient method.

The starting conformations of the dipeptides included both cis and trans isomers. However, preliminary calculations show, as expected, that the cis isomers feature much higher energy than the trans species (by approximatelly 12 kcal/mole). In consequence, rigorous geometric optimization of a number of conformers featuring different initial geometric dealt only with trans isomers.

The psi angle of proline was given initial values of 0°, 60°, 120°, 180°, 240°, and 300°. The same values were used for the phi angle of the second residue (Gly or Ala). A grid method was used combining these values; for example, a starting conformation can feature psi = 0° and phi = 60°. A total of 36 conformations were examined for each peptide, with geometry optimization of all the parameters of the molecule. This procedure ensures the proper identification of the local minima and the global minimum.

Antohi et al. (50) found for the L-Pro-L-Gly dipeptide only five energy-minimized conformations. However, for the L-Ala and D-Ala containing species, 12 and 9 local minima were found, respectively. It is clear that the presence of the methyl increases the number of local minima. There are two gauche local minima for phi (Ala) for every trans local minimum, while in two values of psi of the three conformations of Pro-Gly for which trans minima were observed, no gauche optimized conformations were found.

The lowest-energy conformations for L-Pro-L-Ala and L-Pro-D-Ala are those with the Ala phi close to 180° (−158.4° and 162.8°, respectively). However, while the L-Ala containing dipeptide features the Pro psi angle near −20°, the D-Ala dipeptide has this angle positive for the most stable configurations. The second-lowest-energy conformation for L-Pro-L-Ala exhibits an almost perfect type I beta turn. In contrast, L-Pro-D-Ala has a low-energy conformation that features psi and phi values close to those of a type II beta turn. As far as L-Pro-L-Ala is concerned, its lowest conformation remotely resembling a conformation with a type II beta turn is higher by 13.8 kcal/mole than the global minimum, even higher than the cis structure. It thus appears that while the L-Pro-D-Ala dipeptide could adopt a type II beta turn, it would be practically impossible for L-Pro-L-Ala to do so.

A question that remains to be answered is why the L-Ala species has more optimized structures than the D-Ala species. This fact might have to do with the greater flexibility of the molecule. Indeed, the tridecapeptide with the L-Ala9 residue has not been found in solution to exhibit a structured conformation, while the D-Ala9 containing species exhibits a solution structure stable enough to be studied by NMR (50). Antohi et al. show thus that inferences about the relationship between the biological activity of alpha-factor analogues and their conformations can be supported by ab initio calculations.

Some trends in the values of certain geometrical parameters of the dipeptides investigated in the study (50) are in very good agreement with the experimental data, especially the N(Pro)-Calpha(Pro)-C(Pro) and the

Calpha(Pro)-C(Pro)-O(Pro) angles, which, even though positioned away from the methyl side chain, are affected by the chirality of the alanine residue. In agreement with X-ray crystallographic results, the first angle is smaller for the D-Ala species and larger for the L-Ala species, while the second exhibits the opposite behavior. The L-Pro-D-Ala dipeptide exhibits a conformation very close to that featuring a type II beta turn, as determined by X-ray experiments (50). Since this dipeptide is ten times more active than the L-Ala species, the earlier conclusion that a type II beta turn is necessary for the biological activity of the alpha factor is valid. At the same time, it confirms that a type I beta turn does not ensure activity.

In order to refine their results, Antohi and Sapse (51) considered the blocked dipeptides Ac-L-Pro-L-Ala-NH$_2$ and Ac-L-Pro-D-Ala-NH$_2$. Adding the Ac and NH$_2$ groups allows the formation of a complete beta turn, providing the necessary elements for a ten-membered ring. By computing ab initio energies and geometric parameters of the conformers of these structures, there can be obtained a direct quantum-chemical evaluation of the role of the chirality of the alanyl side chain on the stabilization or destabilization of the bent conformations.

Antohi and Sapse (51) used the Gaussian-90 computer program to solve the Hartree–Fock equations for the Ac-L-Pro-L-Ala-NH$_2$ and Ac-L-Pro-D-Ala-NH$_2$ dipeptides. The calculations were performed using the ab initio method with the 6-31G basis set. The 6-31G is known to predict reliable bond lengths. The addition of polarization functions increases the accuracy of the angles, but the predicted bond lengths are somewhat short. This is corrected by involving the correlation energy in the optimization of the geometry, with, for example, the MP2/6-31G** method (Moller–Plesset of second order). This method requires a large computational effort for as large a system as the blocked dipeptides containing proline. Moreover, Beachy et al. (52) showed that the change in optimum geometry from that predicted by HF/6-31G** to that predicted by MP2/6-31G** is very small. Frey et al. (53) showed that in certain cases the the MP2 optimized geometry is not closer to the experimental energy. Taking this into consideration, Antohi and Sapse (51) considered it sufficient to use the 6-31G basis set at Hartree–Fock level, especially since the main purpose of the work is to compare different conformations to each other. It was thought that the lack of polarization functions and of correlation energy terms would not influence the trends.

The trans peptides were considered. A grid scan procedure similar to the one used in (50) was used. However, in this case the grid is three-dimensional, involving the psi angle of proline, the phi angle of alanine, and the psi angle of alanine. The initial values given to these angles are 0°, 90°, 180°, and 270°. For example, a starting conformation can feature the Pro psi = 0, the Ala phi = 0, and the Ala psi = 90.0°. A total of 64 conformations were examined for each peptide with optimization of all the molecule's parameters. The optimization procedure was the gradient method.

The lowest energy found for Ac-L-Pro-D-Ala was that with psi (Pro) = 103.2°, phi (Ala) = 126.2°, and psi (Ala) = −9.8°C. For Ac-L-Pro-L-Ala, the lowest energy is given by the conformation featuring psi (Pro) = 70.7°, phi (Ala) = −117.7°, and psi (Ala) = 19.5°. As can be seen, in contrast to the nonblocked peptides, where the lowest energies correspond to the extended dipeptides, the presence of the blocking termini accentuates the difference between the two optical isomers. This is due to the formation of the hydrogen bonds. The lowest conformation for the D-Ala species exhibits a type II beta turn, with the torsion angles close in value to those of an ideal type II beta turn. The distance between the oxygen of one terminus and that of the hydrogens of the other is found to be 2.2066 A, a value well suited for a strong hydrogen bond. The bond length between the hydrogen and the nitrogen to which it is attached is somewhat longer, confirming the formation of a hydrogen bond. The oxygen involved in the bond is also fairly close (2.61 A) to the hydrogen of the peptide bond, also forming a seven-membered ring in addition to the 10-membered ring of the beta turn. Probably, the presence of the two rings affords the conformation its high stability. This structure is not only the lowest in energy among the D-Ala containing species, but it is also lower by 1.82 kcal/mole than the lowest conformation of the L-Ala containing dipeptide and lower by 5.77 kcal/mole than the next D-Ala conformation. The potential existence of a bifurcated hydrogen bond was also observed in some tetrapeptide containing a Pro-Gly fragment (51), and it is possible that the characteristics it imparts to the tridecapeptide alpha factor may be responsible for the fact that only a D-Ala analogue that also features this double ring structure is active.

The second energy conformation of Ac-L-Pro-L-Ala-NH2 exhibits a distorted Type I beta turn.

These results confirm the fact that a type II beta turn might be necessary for the biological activity of alpha-factor analogues. This problem accentuates the fact that ab initio calculations can help elucidate relationships between biological activity and structure.

References

1. Raths, S.K., Naider, F., and Becker, J.M. *Journal of Biol. Chem. 263*, 33, 17333 1988.
2. Naider, F., and Becker, J.M. *CRC Critical Reviews in Biochemistry 21*, 3, 225, 1986.
3. Sprague, G.F., and Thorner, J. In *The Molecular Biology of the Yeast Saccharomyces Cerevisiae*, Broach, J.R., Pringle, J.R., and Jones, E.W. (eds.) 657, Cold Spring Harbor Laboratory Press, 1993.
4. Antohi, O. Ph.D. thesis, City University of New York, 1998.
5. MacKay, V., and Manney, T.R. *Genetics 76*, 273, 1974.
6. Hagen, D.C., McCaffrey, G., and Sprague, G.G. *Proc. Natl. Acad. Sci. USA 83*, 1418, 1986.
7. Godchaux, W., and Zimmerman, W.F. *J. Biol. Chem. 254*, 7874, 1979.

8. Moscona-Amir, E., Henis, Y.I., Yechiel, E., Barenholz, Y., and Sokolovski, M. *Biochemistry 25*, 8118, 1986.

9. Higashijima, T., Fujimura, K., Masui, Y., Sakakibara, S., and Miyazawa, T. *FEBS Lett. 159*, 229, 1983.

10. Jelicks, L.A., Shenbagamurthi, P., Becker, J.M., Naider, F., and Broido, M.S. *Biopolymers 27*, 431, 1988.

11. Hagashijima, T., Masui, Y., Chino, N., Sakakibara, S., Kito, H., and Miyasawa, T. *Eur. J. Biochem. 140*, 163, 1984.

12. Jelicks, L.A., Broido, M.S., Becker, J.M., and Naider, F. *Biochemistry 28*, 4233, 1989.

13. Naider, F., Jelicks, L.A., Becker, J.M., and Broido, M.S. *Biopolymers 28*, 487, 1989.

14. Wakamatsu, K., Okada, A., Suzuki, M., Higashijima, T., Masui, Y., Sakakibara, S., and Miyazawa, T. *Eur. J. Biochem. 154*, 607, 1986.

15. Xue, C.B., Erotou-Bargiota, E., Miller, D., Becker, J.M., and Naider, F. *J. Biol. Chem. 264*, 15, 19161, 1989.

16. Masui, Y., Tanaka, T., Chino, N., Kita, H., and Sakakibara, S. *Biochem. Biophys. Res. Commun. 86*, 982, 1979.

17. Matsui, Y., Chino, N., Sakakibara, S., Tanaka, K., Muramaki, T., and Kita, H. *Biochem. Biophys. Res. Commun. 78*, 534, 1977.

18. Levin, Y., Khare, R.K., Abel, G., Hill, D., Eriotou-Bargiota, E., Becker, J.M., and Naider, F. *Biochemistry 32*, 8199, 1993.

19. Bargiota, E.E., Xue, C.B., Naider, F., and Becker, J.M. *Biochemistry 31*, 551, 1992.

20. Higashimjima, T., Miyazawa, T., Masui, Y., Chino, N., Sakakibara, S., and Kito, H. In *Peptide Chemistry*, Yonehara, H., ed. Protein Research Foundation, Osaka, 155, 1980.

21. Garcia-Echeverria, C., Siligardi, G., Mascagani, P., Gibbons, W., Girald, E., and Pons, M. *Biopolymers 31*, 835, 1991.

22. Gounarides, J.S., Xue, C.B., Becker, J.M., and Naider, F. *Biopolymers 34*, 709, 1994.

23. Weiner, S.J., Kollman, P.A., Nguyen, D.T., and Case, D.A. *J. Comp. Chem. 7*, 230, 1986.

24. Press, W.H., Flannery, B.P., Teulolsky, S.A., and Veterling, W.T. In *Numerical Recipies in C, The Art of Scientific Computing*, Cambridge University Press, New York, 301, 1988.

25. Garbow, J.R., Breslav, M., Antohi, O., and Naider, F. *Biochemistry 33*, 10094, 1994.

26. Gullion, T., and Schaefer, J. *J. Mag. Reson. 81*, 196, 1989.

27. Gullion, T., and Schaefer, J. *Adv. Mag. Reson. 13*, 57, 1989.

28. Garbow, J.R., and McWherter, C.A. *J. Am. Chem. Soc. 115*, 238, 1993.

29. Hing, A.W., and Schaefer, J. *Biochemistry 32*, 7593, 1993.

30. Marepalli, H.R., Antohi, O., Becker, J.M., and Naider, F. *J. Am. Chem. Soc. 118*, 28, 6531, 1996.

31. Bitar, K.G., Somogyvari-Vigh, A., and Coy, D.H. *Peptides 15*, 461, 1994.

32. Farlie, D.P., Abbenate, G., and March, D.R. *Curr. Med. Chem. 2*, 654, 1995.

33. Arttamangkul, S., Murray, T.F., Delander, G.E., and Aldrich, J.V. *J. Med. Chem. 38*, 2410, 1995.

34. Bankowsli, K., Manning, M., Seto, J., Haldar, J., and Sawyer, W.H. *Int. J. Pep. Res. 16*, 382, 1980.

35. Sawyer, T.K., Hruby, V.J., Darman, P.A., and Hadley, M.E. *Proc. Natl. Acad. Sci. USA 79*, 1751, 1982.
36. Mierke, D.F., Schiller, P.W., and Goodman, M. *Biopolymers 29*, 943, 1990.
37. Fry, D.C., Madison, V.S., Greeley, D.N., Felix, A.M., Heimer, E.P., Frohman, L., Campbell, R.M., Mowels, T.F., Toome, V., and Wegrzynski, B.B. *Biopolymers 32*, 649, 1992.
38. Kapurniotu, A., and Taylor, J.W. *J. Med. Chem. 38*, 836, 1995.
39. Charpenteir, B., Dor, A., England, P., Pham, H., Durieux, C., and Roques, B.P. *J. Med. Chem. 32*, 1184, 1989.
40. Ning, Q., Ripoll, D.R., Szewczuk, Z., Konishi, Y., and Ni, F. *Biopolymers 34*, 1125, 1994.
41. Yang, W., McKinney, A., Becker, J.W., and Naider, F. *Biochemistry 34*, 1308, 1995.
42. Marepalli, H.R., Yang, W., Joshua, H., Becker, J.M., and Naider, F. *Int. J. Pep. Prot. Res. 45*, 418, 1995.
43. Houston, M.E.Jr., Gannon, C.L., Kay, C.M., and Hodges, R.S. *J. Pept. Sci. 1*, 274, 1995.
44. Antohi, O., Marepalli, H.R., Yang, W., Becker, J.M., and Naider, F. *Biopolymers 45*, 21, 1998.
45. Shenbagamurthi, P., Kundu, B., Raths, S.K., Becker, J.M., and Naider, F. *Biochemistry 24*, 7070, 1985.
46. Naider, F., Gounarides, J., Xue, C.B., Bargiota, E., and Becker, J.M. *Biopolymers 32*, 335, 1992.
47. Gounarides, J.S., Broido, M.S., Becker, J.M., and Naider, F.R. *Biochemistry 32*, 908, 1993.
48. Gounarides, J.S. Ph.D. thesis, City University of New York, 1993.
49. Richardson, J.S., and Richardson, D.C. In *Prediction of Protein Structure and the Principles of Protein Conformation*, Plenum Press, Fasman, G.D., ed. New York, 1990.
50. Antohi, O., Naider, F., and Sapse, A.M. *THEOCHEM 360*, 99, 1996.
51. Antohi, O., and Sapse, A.M. *THEOCHEM 430*, 247, 1998.
52. Beachy, M.D., Chasman, D., Murphy, R.B., Halgren, T.A., and Friesner, R.A. *J. Am. Chem. Soc. 119*, 5908, 1996.
53. Frey, R.F., Coffin, J., Newton, S.Q., Ramek, M., Cheng, V.K.W., Momany, F.A., and Schafer, L. *J. Am. Chem. Soc. 114*, 5369, 1992.

10
Tight Turns in Proteins

Polypeptides often exhibit folded conformations in which the chain reverses its direction over a few residues. Many proteins contain such sequences. In addition to proteins, such conformations have been found in oligopeptide hormones (see Chapter 9) and antibiotics, such as gramicidin SA. Reverse turns, also called tight turns, together with helices and beta sheets, elements of the secondary structure of proteins, control their tertiary structure by their variations. The tight turns are the most prevalent type of nonrepetitive structure recognized. As opposed to helices and beta structures, which have the property that successive residues feature similar ϕ and ψ angles, the nonrepetitive fragments of structure have different ϕ and ψ angles for each residue, so the residue position within the structure influences it more than in a repetitive structure (1).

The most important reverse turns are the beta turns, which comprise about 25% of the residues of proteins (2). These turns occur mainly between two antiparallel beta strands. They thus reverse the direction of the chain (3,4). The beta turns comprise a fragment consisting of a tetrapeptide, with amino acid residues called $i, i + 1, i + 2$, and $i + 3$ (5). Each type of beta turn is characterized by the values of the torsion angles of the $i + 1$ and $i + 2$ residues. The i and $i + 3$ residues usually have beta strand main chain conformation. Venkatachalam (5), who first recognized tight turns from a theoretical conformational analysis, found three general types. The type I beta turn features approximately $\phi_2 = -60°$, $\psi_2 = -30°$, $\phi = -90°$, and $\psi_3 = 0°$. The type II beta turn features $\phi_2 = -60°$, $\psi_2 = 120°$, $\phi_3 = 90°$, and $\psi_3 = 0°$. These types are related to one another by a 180° flip of the peptide unit situated in the center. A third type, the type III beta turn, has repeating ϕ and ψ values of $-60°$ and $-30°$, and thus is identical to the 3_{10}-helix. Types I and III are identical for the second residue and differ for the angles of the third residue by 30°.

Additional types, called I′ and II′ are the mirror images of the type I and II beta turns. Beta turns of types II, II′, and I′ feature dihedral angles with values especially suited to the presence of glycine in a central position. Indeed, glycine, being the smallest of the amino acids, relieves some of the steric strain that might be present in these turns if the central amino acid were larger. Chou and Fasman (6) identified 459 tight turns in proteins and found that 61% of the type II turns feature glycine in position 3 (1). Type II' turns strongly prefer glycine in position 2.

In order to account for all the observed cases, Lewis et al. (7) define five additional types of beta turns, where the alpha carbon of the nth and $n + 3$rd residues are less than 7 A apart (1). The type V beta turn features $\phi_2 = -80°$, $\psi_2 = 80°$, $\phi_3 = 80°$, and $\psi_3 = -80°$. Type V' is its mirror image. The type IV beta turn is a category of turn containing more than one conformation, with dihedral angles more than 40° apart from the other types. Type VI is characterized by a cis proline in the third position. The type VII beta turn has either ϕ_3 near 180° and $\psi_2 < 60°$ or $\phi_3 < 60°$ and ψ_2 near 180°.

It can be noticed (1) that types I and III form a single tight cluster, and their ideal ϕ and ψ values are very close. Richardson (1) suggests that the type III beta turn should be eliminated as a separate category. Since there are a large number of nonideal type I beta turns, they can be grouped in a type Ib category. She also suggests eliminating the types V and V' as separate categories and characterizing the beta turns by the types I, I', II, II', VIa, VIb, and IV, with the possible addition of type Ib. Types VIa and VIb are distinguishable from each other by the fact that VIa has a concave orientation of the middle peptide, while VIb has a convex orientation.

A type II beta turns involves a linear hydrogen bond, and it is approximately planar. In contrast, a type I beta turn is nonplanar, even for ideal cases, and the NH and CO of the hydrogen bond are almost perpendicular to each other (1).

Originally, according to the initial characterization of Venkatachalam (5), the hydrogen bond was an essential feature of the tight turns. However, numerous tight turn conformations were found to possess the putative atoms of hydrogen bonds outside of hydrogen-bond-like distances (8). It was concluded that many beta turns are stable enough not to require stabilization by a hydrogen bond. As pointed out by Richardson (1), this is not surprising, since the turn types consist of α, β, and left-handed glycine conformations combined in different ways. These turns occur mostly at the surface of the protein so there can be hydrogen bonding to the solvent.

The tight turns feature certain residues as those most often encountered. For instance, it was shown (9–10) that many residues are hydrophilic, probably due either to inherent conformational preferences or to the location of the turns near the surface, where they join together internal segments of the secondary structure or interrupt them (1). Energy calculations show glycine to be particularly favorable (6). However, they are not found as commonly as suggested by the energetics. A residue particularly common is proline, not only in the cis-proline turn but also in types I, II, and III turns (1).

Nemethy and Printz (11) discovered another type of tight turn, involving three residues instead of four, as is typical of beta turns. They refer to this type of turn as the gamma turn. It can also be described as a 1:3 turn, as opposed to the beta turns, which are 1:4 turns.

Examples of gamma turns found in proteins comprise the Ser25-Thr26-Tyr27 sequence of thermolysin (12) and the Ile79-Thr80-Val81 sequence found in the satellite tobacco necrosis virus protein (13). In both cases, a 1:3 hydrogen bond is present between the carbonyl oxygen of the first residue and the amino hydrogen of the third residue.

Gamma turns have been observed in small cyclic peptides that contain proline, as, for instance, cyclo-Pro-Gly-Pro-Gly-Pro-Gly (14), which features a Pro-Gly-Pro turn; cyclo-Gly-D-Ala-Pro-Gly-Pro, which features a D-Ala-Pro-Gly turn (15); and the D-Phe-Pro-Gly turn of cyclo-Gly-D-Phe-Pro-Gly-Pro pseudopeptide in solution (16). In each of these cases the central residue is proline.

The C7 structure that features a seven-atom ring formed via a hydrogen bond between CO and NH, which is the 1:3 bond of a gamma turn, has been found experimentally and also predicted theoretically for other systems. Avignon et al. (17) consider the general formula $CH_3CO_3NR_1CHR_2ON$ (R3) 2, which can feature a C5 or a C7 structure, depending on the nature of R1, R2, or R3. Their infrared studies prove the stability of the hydrogen bond.

Nemethy and Printz (11) performed molecular-mechanics calculations to investigate the gamma turn. They defined the conformations in terms of dihedral angles and used for bond lengths and angles standard values. They used a Lennard–Jones 6–12 potential with constants given by Scheraga (18). Hydrogen atoms were considered separately, except for the terminal methyl groups, which were treated as extended atoms (11). The electrostatic interactions were computed in terms of Coulomb interactions between partial charges on each atom, using an effective dielectric constant of 3.5. Calculating the strength of the hydrogen bond, Nemethy and Printz (11) used constants chosen to give a minimum of –5.5 kcal/mole for a linear N–H ... O=C bond with a H ... O distance of 1.85 A. Solvent effects were not taken into consideration.

Values of 0.6, 0.2 and 2.8 kcal/mole were assigned to the rotational barriers about the $N–C^\alpha$, $C^\alpha–C$ and $C^\alpha–C^\beta$ bonds. Limited internal rotation was allowed around the peptide $C'–N$ bond. Nemethy and Printz (11) assumed a simple sinusoidal dependence of the torsional energy on the angle ω. This was done in order to keep the number of variables limited. The torsional energy thus takes the form

$$U(\omega) = \frac{U_0}{2}(1 - \cos 2\omega) - \frac{U_1}{2}(1 - \cos \omega),$$

where $U_0 = 19.0$ and $U_1 = 2.0$ kcal/mole. The energy difference between the trans and cis peptides is in this case 2.0 kcal/mole, and the barrier to rotation as measured from the trans form is 20 kcal/mole (11). No free amino or carboxy groups were considered.

It was found (11) that the dihedral angles of the Ala and Gly sequences do not differ significantly. Moreover, substitutions of other amino acids or incorporation of the tripeptide in longer sequences changes the dihedral angles only slightly. A polypeptide chain folded in a gamma turn can exhibit two hydrogen bonds: one as part of a seven-membered ring and the other as part of a larger ring. When glycyl and alanyl residues were compared, it was found (11) that an L-alanyl chain can be placed into positions 1 or 3 without any strain on the gamma turn, and in position 2 without much strain. However, a D residue cannot be placed in the 1 or 3 position without affecting greatly the turn, and in the 2 position it can be placed with some energy cost. The gamma turn is characterized by the fact that L side chains extend on one side of the backbone turn, being thus in proximity (11). Larger side chains exhibit favorable nonbonded interactions that stabilize the structure.

As pointed out by Nemethy and Printz (11), some rotation around the C'–N bonds is necessary for the stabilization of the hydrogen bonds in gamma turns. The energy necessary for this rotation is small, only 0.9 kcal/mole, and it is compensated by the hydrogen bond stabilization.

As discussed in detail in Chapter 6, the gamma turn was found present in N-formylproline amide and N-acetylproline amide, as predicted by ab initio calculations (Ref. 28, Chapter 6).

When the gamma turn is compared to beta turns, it is evident that the former requires one fewer residue than the latter. Both turns represent a bend in the polypeptide, which folds it in an antiparallel beta structure. As calculated by Nemethy and Printz (11), the intramolecular potential energy per residue is approximately the same for the two types of turns.

The fact that the gamma turn requires one fewer residue than the beta turn has the result that in the hairpinlike structure of the gamma turns there is one more hydrogen bond per residue than in the beta turns.

Another difference between the beta and gamma turns is that in the former the type I beta turn prefers glycine or a D-aminoacid as the $i + 1$ residue, and the same is true for the type II beta turn for the $i + 2$ residue. If in these positions there is an L-alanine residue, the energy is increased by 5 or 8 kcal/mole, respectively (see also Chapter 9 in this respect). In contrast, in the gamma turn, the position $i - 1$ can be occupied by either an L or a D residue, with energies of only 2 or 2.5 kcal/mole above glycine (11).

The hydrogen bonds in beta turns are linear, and thus are as strong as possible. However, they are not shielded well from the solvent by the side chains. This situation also arises in the gamma turn for the linear i to $i + 2$ hydrogen bond. The $i + 2$ to i hydrogen bond, though, is bent, and is therefore weaker, but it is more shielded from the solvent by the side chains and backbone (11). Therefore, both structure feature a reduction of amide–water proton exchange rates.

One of the methods of distinguishing between beta and gamma turns is NMR, more specifically, the determination of the N–C coupling constant.

Möhle et al. (19) point out the fact that reverse turns assist in the consolidation of the compact structure in globular proteins. The turns may play an important role in recognition problems (20–23) and in general in protein folding (24–26). To understand these phenomena better, a knowledge of the structure and energetics of beta turns is necessary. Since the AM1 semiempirical method is not capable of locating the beta I turn, ab initio methods are required for an exact description of tight turns. Perczel et al. (27) applied ab initio calculations with the 3-21G basis set to the study of beta turns. Bohm (28) examined the type I and type II beta turns of Ac-Gly-Gly-NHCH3, using ab initio calculations with the 3-21G basis set and higher basis sets, also performing MP2 single-point correlation energy calculations.

Möhle et al. (19) undertook a systematic study of beta turns using ab initio calculations with basis sets of different sizes. They also estimated correlation energy effects, thermodynamic properties, and solvent effects on the structures of the peptides.

They also compared beta turns with glycine to those with alanine in order to examine the influence of the specific residues on the structure of the turn. Mohle et al. (19) considered several peptides, such as Ac-Gly-Gly-NHMe and Ac-L-Ala-L-Ala-NHMe. They used a general formula: CH3CO-X -Y -NHCH3, where X can be Gly, L-Ala, or Aib (alpha-aminoisobutyric acid), and Y can be Gly, L-Ala, D-Ala, Aib or L-Pro. These types of sequences serve as prototypes to embody the principal structural characteristics of beta turns.

Möhle et al. (19) used the 6-31G* and the 3-21G basis sets to perform the ab initio calculations, with complete geometric optimization. The peptides with X = Gly and Y = Gly were optimized at MP2/6-31G* level, which is a high level for a system of this size. For this molecule, the calculations predict that at MP2/6-31G* level, HF/6-31G* level, and HF/3-21G level, the type II beta turn is more stable than the type I beta turn. As reported by Möhle et al. (19), the preference of type II over type I is clear in turns with glycine, a fact that leads to beta turns of type II with glycine in the $i + 2$ position being termed "glycine turns." When Möhle et al. (19) calculated the energy of a glycine turn for X = L-Ala and Y = Gly, using the HF/6-31G* method, they found it more stable by 6.2 kJ/mole than the type I beta turn. The difference in energy between a type I and a type II beta turn is somewhat dependent on the basis set used. For the HF/3-21G method, Möhle et al. obtained 1.4 kJ/mole, while the MP2/6-31G* method predicts a difference of 3.4 kJ/mole. Calculations performed with the STO-3G minimal basis set for N-acetylalanylglycine amide (see Chapter 11) predict the type I beta turn to be more stable than the type II beta turn. Entropy and solvent effect calculations are necessary to reverse the trend.

As reported by Möhle et al. (19), the extended conformation is more stable than the turn-featuring conformations at HF/6-31G* calculational level. However, when the correlation energy is considered, the order of stability is reversed. This is explainable by the fact that the cyclic structures are favored by the correlation energy effects. A similar effect is seen for the comparison of C5 and C7 rings. Indeed, the C5 ring is more similar to an extended conformation, with $\phi \approx -160.0°$ and $\psi \approx 160.0°$. At HF/6-31G* calculational level, the C7 ring is more stable by 1.0 kJ/mole, while at MP2/6-31G* level the difference in energy increases to 6.5 kJ/mole (29). However, when the zero-point vibrational energies and other thermodynamical effects are considered, the Hartree–Fock stability order is reestablished (19). Indeed, the entropy effect stabilizes the extended form, as opposed to the cyclic ones, which are more restricted. Due to the cancellations of the correlation effects on one hand and of the thermodynamical factors on the other, the Hartree–Fock/6-31G* results can be considered reliable.

Möhle et al. (19) showed that the torsion angles of the turns, as obtained at different calculational levels, are quite close. The MP2/6-31G* and the HF/6-31G* angles are similar, and the HF/3-21G angles are within acceptable range. As they point out, in the peptide with X = Gly and Y = Gly, the torsional angles agree well with the standard values, but there are some differences. For example, the ϕ angle of the $i + 2$ residue in type I beta turns is closer to $-100°$ than to $-90°$. The ψ angle of the same residue is not zero, but closer to $20°$ for type I and $-20°$ for type II. These values are similar to the values in some globular proteins (30).

As reported by Möhle et al. (19), the main observed difference in alanine-containing beta turns is the fact that type I and type I', as well as type II and type II' beta turns are not equivalent. The stability order resulting from the calculations is I > II > II' > I'. This result was obtained at all calculational levels (19). The correlation energy effects stabilize the cyclic conformations vis-à-vis the extended one, in the same way as in the glycine case, and the same cancellation with entropy effects appears.

The results reported by Möhle et al. (19) agree with the results of Yan et al. (31), who study this kind of system in aqueous solutions with molecular-dynamic methods.

When the D-Ala-containing peptide is studied (19), the stability order, as expected, is I' > II' > II > I, and the torsion angles are simply the negatives of the torsion angles obtained for the L isomers. It results, as pointed out by Möhle et al., that the replacement of two consecutive L-enantiomers that feature type I as their most stable conformation by their D counterparts would make type I' the most stable.

If only one L enantiomer is replaced by D, namely the $i + 2$ residue, Möhle et al. (19) show that the type II beta turn is the most stable. This is in agreement with the studies on the alpha factor discussed in Chapter 9. It is also in agreement with the molecular-dynamic calculations (31). It can thus be seen that in order to synthesize a peptide featuring a type II beta turn, it is advisable to use a D enantiomer. If the precedent residue is replaced by a D enantiomer, the stability order becomes II' > I > I' > II.

If the $i + 2$ residue is Aib, mostly type II beta turns are obtained. For structures with X = L-Ala and Y = Aib, an order of stability of II > I > I' > II' is reported by Möhle et al. (19). Since the increased stability of the type II beta turn over the type I beta turn is also reported for glycine, it can be deduced that both the absence of side chains and a symmetrical disubstitution at the C of the $i + 2$ residue favors the type II beta turn (19). When Aib is inserted as the $i + 1$ residue, there is comparable stability between the type I and the type II' beta turns (19). When this stability is compared to that of the extended conformation, Möhle et al. found that there is an increased tendency to turn formation in peptides with disubstituted amino acids.

When the $i + 2$ residue is L-proline, beta turns with cis peptide bonds can occur. These turns are termed type VIa and VIb. The calculations of Möhle et al. (19) indicate that the VIa type is more stable and that both cis turns are more stable than the type I beta turn with a trans peptide bond.

Solvent effect has an important influence on the order of stability of different types of beta turns. As is shown in Chapter 11, the consideration of solvent effects, coupled with entropy effects, reverses the order of stability of tight turns of N-acetylalanyl glycine amide.

Möhle et al. (19) use the SCRF and the PCM methods for the peptides with X = Gly, Y = Gly; X = L-Ala, Y = L-Ala; and X = L-Ala, Y = L-Pro.

The SCRF method allows for complete geometry optimization. Both methods contain a series of approximations, so the results have to be examined more in the sense of trends than in a numerical, quantitative, way.

The type I beta turn is stabilized by the solvent, in preference to the other types of beta turns. In the cases where the type I beta turn was not the most stable, it becomes so, and in the cases where it was the most stable it does so by a larger energy difference. The reason for this additional stability of the type I beta turn is the fact that it has a higher dipole moment than other types of turns. Another study (32) confirms the stabilization of type I beta turns by the solvent. More possible explanations about this fact are discussed in Chapter 11.

Möhle et al. (19) point out that the quantum-chemical results describe well the dependence of the stability of beta turns of different types on the amino acid residues present. In order to investigate this problem better, they perform molecular-dynamics calculations on the systems with two L-Ala residues, with one L-Ala and one D-Ala residue, with one L-Ala residue and an Aib residue in $i + 1$ and the $i + 2$ positions, respectively, and vice versa. The results indicate that the first system prefers a type I beta turn, while the second prefers a type II beta turn. The other systems feature a mixture of type II, type II' and type II beta turns of similar energies.

The above-mentioned study (19) concludes that the stability of various types of beta turns is strongly influenced by the amino acid residues that make up the given sequence. Quantum-chemical and molecular-mechanics methods are a powerful tool for the design of beta turns.

Deslauriers et al. (30, Chapter 11), report the studies of conformational properties of cyclo L-alanylglycyl-ε-aminocaproyl (cyclo-L-Ala-Gly-Aca). Indeed, cyclo-Ala-Gly-Aca is particularly suitable for the studying of the behavior of oligopeptides chains featuring turns because it involves steric constraints. This molecule must exist in a bent conformation because of the steric constraints introduced by the $(CH2)5$ chain.

Deslauriers et al. confirm the prediction of theoretical studies on this molecule, predicting that it exists mostly in a type II beta turn, with some type I and type III beta turns mixed in. They perform NMR and circular dichroism studies, as well as infrared and Raman studies (33).

The results of the NMR study indicate that cyclo L-Ala-Gly-Aca features a well-defined conformation. The hydrogens of the Aca residue give NMR signals typical of a cyclohexane conformation and not of a hydrocarbon chain. The coupling constants are consistent with rigid staggered arrangements about the $C^\alpha-C^\beta$ and the $C^\delta-C^\epsilon$ bonds. The vicinal coupling of the hydrogen set on C^α with the hydrogens set on the two C^β's indicate a trans and gauche position, respectively. These and other observations (Ref. 30, Chapter 11) are compatible with a beta turn, featuring a bent hydrogen bond between the Aca NH and the Aca C=O. The experimental data suggest a model with planar trans peptide bonds and angles of $(\phi, \psi)_{Ala} = (80°, 70°)$, $(\phi, \psi)_{Gly} = (120°, 20°)$, and almost staggered arrangements around the $C^\alpha-C^\beta$ and the $C^\delta-C^\epsilon$ bonds. This conformation corresponds to a type II beta turn. This result is in agreement with the computations of Nemethy et al. (34).

Circular dichroism experiments suggest that cyclo L-Ala-L-Ala-Aca features mostly type I and III beta turns, while cyclo L-Ala-D-Ala-Aca features mostly type II beta turns (Ref. 30, Chapter 11). It was also found that cyclo L-Ala-Gly-Aca contains about 85% type II beta turn in trifluoromethanol, 80% in methanol, and 65% in water. The calculations of Woody (35) showed that the circular dichroism method is very sensitive to variation in those conformations that are near the Venkatachalam beta turn conformations.

Experiments on Ac-L-Ala-Gly-NHCH$_3$ and Boc-L-Ala-Gly-Aca-OMe (Ref. 30, Chapter 11) show a preference of the former for a type II beta turn. The second also shows a great probability of exhibiting a type II beta turn.

The authors of the study conclude that the predominant conformation of cyclo L-Ala-Gly-Aca in solution is a type II beta turn. As shown by Nemethy et al. (34), the type I beta turn conformation is higher in energy by 0.74 kcal/mole. Thus, the type II beta turn conformation will exist in 63% of the molecules at 25°C (as indicated by the Boltzmann distribution).

The computational studies (34) do not take into consideration the solvent interaction. As shown before and discussed further in Chapter 11, the solvent lowers even more the energy of type II beta turns.

Sathyanarayana and Applequist (36) studied the theoretical absorption and circular dichroism spectra of beta turn model peptides. Among the peptides studied are Ac-Gly-Gly-NHMe, Ac-L-Ala-L-Ala-NHMe, and Ac-L-Ala-Gly-NHMe in the type I, II, and II' beta turn conformations.

In addition, they study cyclo L-Ala-Cly-ε-aminocaproyl, cyclo L-Ala-L-Ala-ε-aminocaproyl, and cyclo L-Ala-D-Ala-ε-aminocaproyl in their minimum energy conformations.

The circular dichroism spectra of beta turns is described theoretically by Woody (35), but some of the experiments disagree with the computations, leading to the need for further investigation (36).

The study of Sathyanarayana and Applequist uses the dipole interaction theory, which has been successfully used before in a number of studies (37–42). This theory investigates the interaction among induced electric dipole moments of all atoms and NC'O groups in the molecule. The calculated absorption and the CD spectra are due to the transition in the NC'O group at 190 nm. Indeed, only the 190 nm transition is treated as dispersive. The atom coordinates were obtained from known bond lengths, bond angles, and dihedral angles, as done by Ramachandran and Sasisekharan (43). For the methyl groups in Ala side chains staggered conformations were used, while for the methyl groups in acetyl and methyamide ends, semieclipsed conformations were used.

The structural variables were the backbone dihedral angles, which in the case of Ac-X-Y-NHMe molecules were given the standard beta turn values, with planar trans conformation for the peptide bond. The peptides containing aminocaproyl, the angles corresponding to the lowest-energy states, as obtained by Nemethy et al. (44) were used.

While Deslauriers et al. (Ref. 30, Chapter 11) found mostly a type II beta turn conformation to characterize Ac-L-Ala-Gly-NHMe, Sathynarayana and Applequist (36) found for this molecule a mixture of different types of beta turn, with a slight predominance of type II. For cyclo L-Ala-Gly-Aca, Deslauriers et al. found mostly a type II beta turn, in agreement with the results reported in (36).

For cyclo L-Ala-L-Ala-Aca, the results of Bandekar et al. (45) are in disagreement with those of Sathynarayana and Applequist (36), and so are the results reported for cyclo L-Ala-D-Ala-Aca.

It might be concluded that high-level ab initio calculations are more reliable in the prediction of different types of tight turns, and as the computational capacity for performing such calculations increases, larger systems will be investigated.

References

1. Richardson, J.S. In *Advances in Protein Chemistry*, Vol 34, Anfinsen, C.B., Edsall, J.T., and Richards, F.M., (eds.) Academic Press, New York, 1981.
2. Kabsch, W., and Sanders, C. *Biopolymers 22*, 2577, 1983.

3. Milner-White, E.J., and Poet, R. *Trends Biochem. Sci. 12*, 189, 1987.
4. Sibanda, B.L., and Thornton, J.M. *Nature 316*, 170, 1985.
5. Venkatachalam, C.M. *Biopolymers 6*, 1425, 1968.
6. Chou, P.Y., and Fasman, G.D. *J. Mol. Biol. 115*, 135, 1977.
7. Lewis, P.N., Momany, F.A., and Scheraga, H.A. *Biochem. Biophys. Acta. 303*, 211, 1973.
8. Crawford, J.L., Lipscomb, W.N., and Schellman, C.G. *Proc. Natl. Acad. Sci.* USA *70*, 538, 1973.
9. Kuntz, I.D. *J. Am. Chem. Soc. 94*, 4009, 1972.
10. Rose, G.D. *Nature 272*, 586, 1978.
11. Nemethy, G., and Printz, M.P. *Macromolecules 6*, 755, 1972.
12. Matthews, B.W. *Macromolecules 6*, 818, 1972.
13. Jones, T.A., and Liljas, L. *J. Mol. Biol. 177*, 735, 1984.
14. Madison, V., Atreyi, M., Deber, C.M., and Blout, E.R. *J. Am. Chem. Soc. 96*, 6725, 1974.
15. Pease, L.G., and Watson, C. *J. Am. Chem. Soc. 100*, 1279, 1978.
16. Spatola, A.F., Anwer, M.K., Rockwell, A.L., and Gierasch, L.M. *J. Am. Chem. Soc. 108*, 825, 1986.
17. Avignon, M., Huong, P.V., Lascombe, J., Lascombe, M., Marraud, M., and Neel, J. *Biopolymers 8*, 69, 1969.
18. Gibson, K.D., and Scheraga, H.A. *Proc. Natl. Acad. Sci.* USA *58*, 420, 1967.
19. Möhle, K., Gusmann, M., and Hofmann, H.J. *J. Comp. Chem. 18*, 1716, 1997.
20. Stanfield, R.L., Fieser, T.M., Lerner, R.A., and Wilson, I.A. *Science 248*, 712, 1990.
21. Marshall, G.R. *Curr. Opin. Struct. Biol. 2*, 904, 1992.
22. Nikiforovich, G., and Marshall, G.R. *Int. J. Pep. Prot. Res. 42*, 171, 1993.
23. Nikiforovich, G., and Marshall, G.R. *Int. J. Pep. Prot. Res. 42*, 181, 1993.
24. Scholnik, J., and Kolinski, A.J. *J. Mol. Biol. 221*, 499, 1991.
25. Soman, K.V., Karini, A., and Case, D.A. *Biopolymers 31*, 1351, 1991.
26. Dill, K.A., Fiebig, K.M., and Chan, H.S. *Proc. Natl. Acad. Sci.* USA *90*, 1942, 1993.
27. Perczel, A., McAllister, M.A., Csaszar, P., and Csizmadia, I.G. *J. Am. Chem. Soc. 115*, 4849, 1993.
28. Bohm, H.J. *J. Am. Chem. Soc. 115*, 6152, 1993.
29. Frey, R.F., Coffin, J., Newton, S.Q., Ramek, M., Cheng, V.K.W., Momany, F.A., and Schafer, L. *J. Am. Chem. Soc. 114*, 5369, 1992.
30. Wilmot, C.M., and Thornton, J.M. *Prot. Eng. 3*, 479, 1990.
31. Yan, Y., Erickson, B.W., and Tropsha, A. *J. Am. Chem. Soc. 117*, 7592, 1995.
32. Osapay, K., Young, W.S., Bashford, C., Brooks III, C.L., and Case, D.A. *J. Phys. Chem. 100*, 2698, 1996.
33. Maxfield, F.R., Bandekar, J., Krimm, S., Evans, D.J., Leach, S.J., Nemethy, G., and Scheraga, H.A. *Macromolecules 14*, 997, 1981.
34. Nemethy, G., McQuie, J.R., Pottle, M.S., and Scheraga, H.A. *Macromolecules 14*, 975, 1981.
35. Woody, R. In *Peptides, Polypeptides and Proteins*. Blout, E.R., Bovey, F.A., Goodman, M., and Lotan, N., (eds.) Wiley, New York, 1974.
36. Sathyanarayana, B.K., and Applequist, J. *Int. J. Pep. Prot. Res. 27*, 86, 1986.
37. Applequist, J. *J. Chem. Phys. 70*, 4332, 1979.
38. Applequist, J. *Biopolymers 20*, 387, 1981.
39. Applequist, J. *Biopolymers 20*, 2311, 1981.

40. Applequist, J. *Biopolymers 21*, 799, 1982.
41. Caldwell, J.W., and Applequist, J. *Biopolymers 23*, 1891, 1984.
42. Sathyanarayana, B.K., and Applequist, J. *Int. J. Pep. Res. 26*, 518, 1985.
43. Ramachandran, G.N., and Sasisekharan, V. *Adv. Protein Chem. 23*, 283, 1968.
44. Nemethy, G., and Printz, M.P. *Macromolecules, 6*, 755, 1972.
45. Bandekar, J., and Kuirm, S. *Int. J. Pept. Prot. Res. 26*, 407, 1985.

11
Some Small Peptides

In order to model biological systems successfully, it is necessary to use an accurate model for calculating the energetics of peptide conformations both in gas phase and in solutions. Knowledge of peptide conformation can provide information about peptide–receptor binding, which, as shown in some of the previous chapters, is the basis of some crucial biological processes. In addition, this knowledge can provide a basis for rational drug design.

Experimental methods employed in the study of peptide conformation include NMR, infrared spectroscopy, and circular dichroism. Theoretical methods include molecular-mechanics force fields and molecular orbital methods, semiempirical, and ab initio. The molecular-mechanics methods comprise AMBER, OPLS, CHARMM, MMFF, and MM# (1). The force fields used by these methods are parametrized to fit experimental data and, sometimes, ab initio results. In both cases, the molecules studied are small, and it is necessary to create analytic potentials applicable to large peptides and proteins.

Beachy et al. (1) have developed novel quantum-chemical methods capable of treating large systems at a high calculational level, including electron correlation energies. Such calculations require the use of supercomputers, which can also help determine the accuracy of force fields for the prediction of peptide energetics.

Beachy et al. (1) studied the alanine tetrapeptide that consists of three alanine residues blocked at the N terminus by an acetyl group and at the C terminus by an N-methylamine group. They generated initial geometries by a limited conformational search in the MacroModel programs (2) using the AMBER force field. They chose the five lowest conformations and five conformations of higher energy. Subsequently, these conformations were geometry optimized with the 6-31G** basis set, at Hartree–Fock level.

MP2 single-point calculations were performed by Beachy et al. (1) at the 6-31G** optimized geometries, with the triple-zeta correlation consistent basis of Dunning (3), without the f orbitals.

Alanine dipeptide calculations (1) were performed using the Dunning basis set, at MP2 level, using the Hartree–Fock 6-31G** optimized geometries. Local minima were located for the MacroModel, MMFF, and Charmm force fields. These geometries differ somewhat from the ab initio derived geometries (1). Comparison between the ab initio and the molecular-mechanics results was also performed by Beachy et al. (1) by carrying out force-field optimizations in which the psi and phi torsional angles were frozen to the values obtained by the 6-31G** calculations. Typically, these angles as calculated by force field methods vary from the 6-31G** torsional angles by 0.1°.

The study reported by Beachy et al. (1) also investigates the accuracy of force-field methods in reproducing hydrogen bonds in beta sheets (1). For this purpose, force-field calculations were performed for antiparallel beta sheets of two alanine dipeptides with torsion angles kept at their ab initio calculated values.

The quantum-chemical calculations were performed with the PSGVB method (1), which saves computational effort.

Murphy et al. (4) and St. Amant et al. (5) have performed extensive investigations of how accurate different methods, such as Hartree–Fock and local and canonical MP2 are for the evaluation of energy differences as compared with experimental values. As shown (4), the local MP2 method with a large basis set such as Dunning's provides the highest level of accuracy, while the same method with a small basis set might lead to qualitative errors. The geometries used for the single-point calculations somewhat influence the errors. For example, the error for HF/6-31G** geometry is 0.42 kcal/mole, while the error for the MP2/6-31G* geometry is 0.36 kcal/mole (1). Murphy et al. (6) show that the value of 1 kcal/mole for the deviation from experiment of the energy of methyl vinyl ether is due to the low-lying (π) state associated with the C=C double bond. However, as shown by Beachy et al. (1), substitution of HF/6-31G** geometries for MP2/6-31G* geometries increases the errors by only a few tenths of a kilocalorie per mole.

As rationalized by Beachy et al. (1), the investigation of the alanine dipeptide (acetyl-Ala-NHCH3) could shed light on the differences between the methods applied, differences that are carried out to the tetrapeptide. Therefore, they compare their results on the molecule with previous results reported in the literature and with results obtained by the Gaussian-94 program. They also investigate the difference between using a local MP2 versus a canonical MP2 calculation. They also examine the differences between using a Hartree–Fock geometry and an MP2 geometry.

Gould et al. (7) have optimized this molecule with the 6-31G** basis set at Hartree–Fock level. Comparing the MP2/6-31G* results for the torsional angles with the results obtained by Gould et al. (7), one notices quite small differences, with a maximum change of 10° (1). In addition to the minima obtained by Gould et al. (7), Head-Gordon et al. (8) find an additional minimum when they study an alanine dipeptide analogue.

The most stable conformer is found to be the C7eq conformer, which features a seven-membered ring. The trend observed for HF/6-31G** geometry holds for the MP2/6-31G* results. This result is in agreement with the calculations of Frey et al. (9), who find a difference of only 0.33 kcal/mole for N-formylalanylamide, when comparing the energy corresponding to HF/6-311G** geometries to that corresponding to MP2/6-311G** geometries.

For alanine tetrapeptide structures, Beachy et al. (1) consider ten conformers, with most of them featuring phi and psi pairs in regions found in protein structures (10). These conformers, as pointed out by the authors of the study, can be viewed as representative of structures found in proteins outside of beta sheets or alpha-helical regions. However, they do not represent the lowest-energy conformers, as found by ab initio calculations (1). The ten conformers range from compact structures with 1 to 3 hydrogen bonds to relatively extended structures.

All dipeptide minima are represented. Several conformers contain C7ax fragments, which are rarely found in protein structures.

One question raised by Beachy et al. (1) is whether their sample is large enough to draw conclusions about the agreement of the force fields and quantum-chemical data. They conclude that ten tetrapeptide conformations representing 30 phi, psi pairs does constitute a system large enough to be meaningful for gross differences between the two methods. They find that the best performance in comparison with the ab initio calculations is obtained with OPLS (11) and MMFF (12,13) methods.

When the energetics of the systems were examined, it was found (1) that the HF calculations predicted lower energies for the extended and partially extended structures as compared to the MP2 results. One explanation for this fact might be (1) that the more compact structures have greater dispersion energy, better described by the inclusion of the correlation energy effects.

Beachy et al. (1) also examined a beta-sheet dimer with two intermolecular carbonyl-amide hydrogen bonds. They compared quantum-chemical and molecular-mechanics energetics for one point on the hypersurface of the complex. They computed the binding energy of the dimer as the difference between the energy of the complex and the sum of the energies of the monomers, correcting the basis set error (BSSE) by the counterpoise method (14). The interaction energies thus calculated vary between −7.23 kcal/mole, as calculated by the MM3 molecular-mechanics method, to −11.68 kcal/mole for noncounterpoise corrected HF/6-31G**, to −17.63, as calculated by the OPLS* method. It is observed, in general, that some of the molecular-mechanics methods used, such as AMBER94, OPLS, and CHARMM, overestimate the binding energy. In the above calculations the monomers were kept rigid. Beachy et al. (1) performed a set of calculations with optimization of the monomers' geometries. In this case, they found for the binding energies −11.80 kcal/mole for the alanine dipeptide, using the

HF/6-31G* calculations. Again, the highest binding energy is predicted by the OPLS* molecular-mechanics method, with a value of −16.44 kcal/mole. It is suggested (1) that force fields that represent electrostatic interactions only in terms of atom-centered partial charges do describe accurately the geometries and energies of hydrogen-bonded systems.

All the above results suggest that a truly accurate quantitative description of peptide and protein energetics in gas phase using the molecular-mechanics methods is not yet feasible. One of the reasons for this fact might be that the parameters have been obtained from structures in the condensed phases. However, Beachey et al. (1) refute this argument on the basis that there is no single condensed-phase environment, but the environment of a protein is different from bulk water. Beachy et al. (1) suggest that high-level correlated quantum-chemical results for intermolecular interactions should be incorporated directly into molecular-mechanics force fields.

Beachy et al. (1) offer the following assessment of the efficiency of molecular-mechanics calculations for the conformational analysis of peptides and proteins: There is a large variety of molecular-mechanics methods in terms of their prediction of geometrical parameters, such as torsional angles. Some of them heavily parametrize the torsional angles, while some of them do not. However, there is no correlation between the range of torsional energies and the range of total conformational energies. A large torsional energy is not necessarily associated with a large conformational energy. It was also suggested (1) that rigorous parametrization with the use of ab initio data of high quality will eventually yield a molecular-mechanics method using a force field capable of describing systems larger than those for which the parameters have been obtained. It is difficult to brand one of the molecular methods as "the best" (1). There are significant differences between them in the ability to locate the ab initio minima on the potential energy surface of the tetrapeptide. However, they suggest (1) that the MMFF (1994) version is among the best.

The ability of a calculational method to correctly predict hydrogen-bond formation is of great importance. Indeed, interchain or intrachain hydrogen bonds account for many structural features of peptides and proteins and play a central role in the secondary structure of proteins.

Formamide has been used as a model for peptides, since it is the smallest molecule possessing a peptidic-like bond. As early as 1968, Pullman et al. (15) studied the structure of formamide with the semiempirical CNDO method. Schlegel et al. (16) and Zielinski and Poirer (17) applied ab initio calculations to the study of formamide, using double-zeta basis sets. Johansson et al. (18) used ab initio calculations with the STO-3G and 4-31G basis sets to assess the relative strength of amide–amide hydrogen bonding as compared to amide–water hydrogen bonding. Scheiner and Kern (19) report a study employing PRDDO (partial retention of diatomic differential overlap) for calculations in amide hydrogen bonding. Hinton and Harpool (20) studied higher oligomers of formamide with the STO-3G basis set.

Sapse et al. (21) applied ab initio calculations to the study of the formamide dimer, with optimization of the monomers, and to the study of N-methylacetamide and its cis–cis dimer. They used the STO-6G, 6-31G, and 6-31G* basis sets for the investigation of formamide, imposing the planarity constraint and the symmetry of the monomers in the dimer. For N-methylacetamide they used the STO-6G and 6-31G basis sets, with the same constraints, except for the methyl group. In addition, a complex formed by acetylamine and N-acetylglycinamide was also studied with the STO-6G basis set. In this complex the two molecules are arranged in a parallel pleated sheet arrangement (21). Another complex studied is that formed by N-methylacetamide and two molecules of formamide, one bound to the hydrogen on the nitrogen, one bound to the oxygen of the carbonyl group. In this complex the N-methylacetamide is in a trans conformation.

In the formamide dimer, the monomers show, as expected, some elongation of the C=O and N–H bonds, a slight increase of the OCN angle, and a slight decrease of the HNC angle. The STO-6G calculations overestimate the CN bond length. There are no significant differences in the other bond lengths, as predicted by the STO-6G, 6-31G, and 6-31G* basis sets. In the dimer, the N–H–O angle takes a value of 165°. As shown by Johansson et al. (18), the energy dependence on an angle over 160° becomes very flat. In N-methylacetamide there is very little difference between the parameters of the cis and trans isomers.

The parallel pleated sheet complex between N-acetylglycinamide and acetylamine features a NHO angle of only 150.9°, a fact that might account for a weaker hydrogen bond.

When the energetics of various isomers were investigated, it was found that the trans isomer of N-methylacetamide is more stable than the cis isomer by 2.6 kcal/mole (6-31G calculations) and 1.42 kcal/mole (STO-6G calculations). The 6-31G is in closer agreement with the experimental result of about 2 kcal/mole (22). In the case of N-acetylglycinamide, the trans isomer is more stable than the cis isomer by about 3 kcal/mole (21).

The complexation energy of formamide was found to be 10 kcal/mole with the STO-6G basis set, 15 kcal/mole with the 6-31G basis set, and 12 kcal/mole with the 6-31G* basis set. There is no doubt that the 6-31G* basis set would provide the best dimerization energies, as compared with minimal basis sets and with basis sets without polarization functions. This fact is partially due to a smaller superposition error. However, it is interesting to notice that the STO-6G results are closer to the 6-31G* results. The 3-21G predicted complexation energy is 20 kcal/mole. This result is obviously too large, and it might be due to a large superposition error. The resemblance between the STO-6G basis set results for the dimerization energy and the 6-31G* results might be due to a cancellation of errors for the former.

Indeed, since the STO-6G minimal basis set must feature a large superposition error, this might be canceled by the fact that the STO-6G basis set predicts a smaller charge separation than double-zeta sets. As such, the two effects work in opposite directions and cancel each other.

The geometry of the hydrogen bonds does not differ significantly from one basis set to another. Since, as mentioned before, over an X–H–N angle of 160° the energy is very flat, small angle changes cannot account for large differences in the binding energy.

The dimerization energy of the N-methylacetamide is found to be slightly lower than that for formamide, a fact that can be explained by the lower charge on the hydrogen (as found via Mulliken population analysis). The complexation energies of the formamide-N-acetylglycineamide complex and of N-acetylglycinamide complex with aminoacetyl are quite small (21).

Sapse et al. (23) applied quantum-chemical ab initio calculations to the conformational analysis of N-acetylalanylglycine amide. This study was part of a project to redesign betabellin (24), a nonbiological protein that contains six beta turns.

As Ac-Ala-Gly-NH2 contains three peptide bonds, it can form either a gamma turn (1;3 hydrogen bond) or a beta turn (1;4 hydrogen bond). It cannot form both, since, as noted by Venkatachalam (25), the two types of turns require different phi and psi backbone dihedral angles.

Sapse et al. (23) used the STO-3G basis set to perform geometry optimization and energy calculations on N-acetylalanylglycine amide. They used the Berny optimization method and examined the second-derivative (Hessian) matrix to make sure that all the eigenvalues were positive so that the structure was a real minimum on the energy hypersurface.

Initial geometries were chosen as follows: The bond lengths and angles were given standard values, while the dihedral angles were chosen in such a way as to describe the type I, type I′, type II, type II′, and type III beta turns, according to the classification proposed by Venkatachalam (25). In addition, structures featuring dihedral angles proper for the double C7 ring, the single C7 ring, and the fully extended structure were also investigated. A few other twisted structures not featuring hydrogen bonds were examined. A total of eleven conformations were examined. Five of them were found to be stable, within 5 kcal/mole of the most stable structure. These structures are shown in order of stability in Figure 11.1 and labeled a, b, c, d, and e. Structure a features two C7 rings with one 1:3 hydrogen bond. Structure b has a type I beta turn with a 1:4 hydrogen bond. Structure C has a type II beta turn with a 1:4 hydrogen bond. Structure d has a single C7 ring with a 1:3 hydrogen bond. Structure e was constrained to adopt a fully extended conformation with no hydrogen bonds. A less extended structure, featuring an extended initial conformation but allowing the dihedral angle to optimize, was termed 5′ (not shown).

The other six structures were found to be higher in energy. The structure featuring a nonclasical C7 ring is higher in energy than structure a by

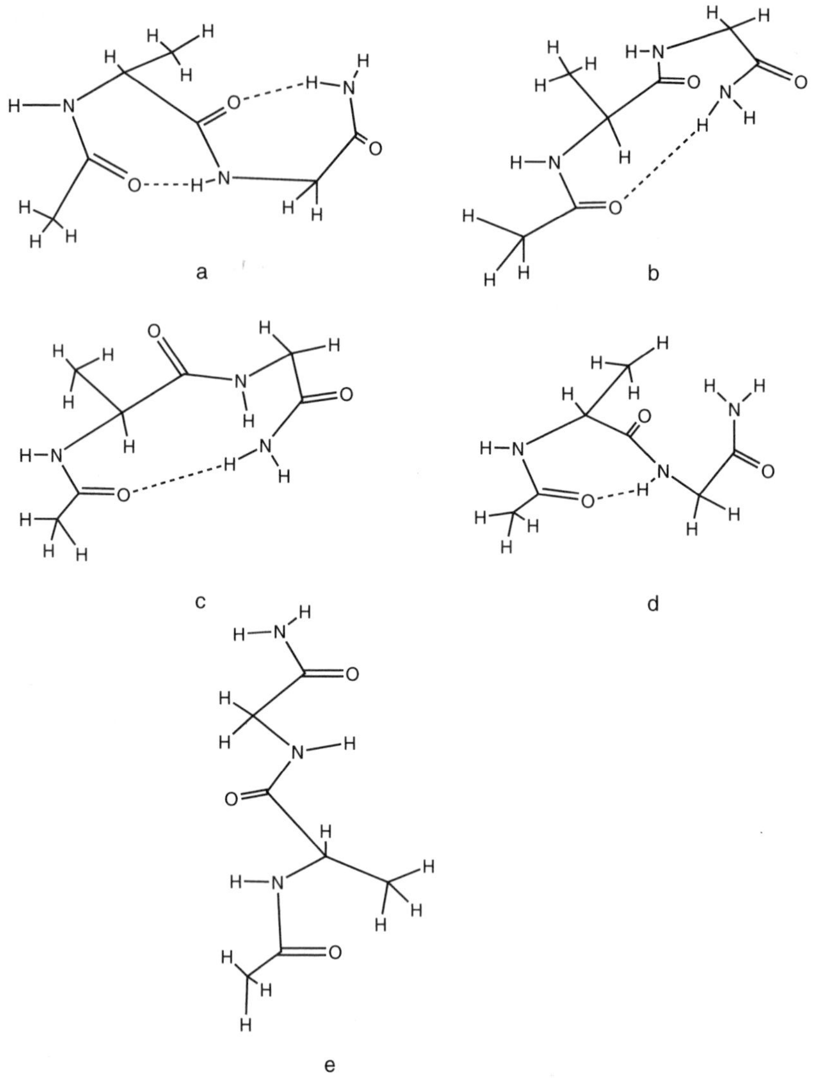

FIGURE 11.1 a. Conformer of N-acetylalanylglycine amide with two 1:3 rings. b. Conformer of N-acetylalanylglycine amide showing a type I beta turn. c. Conformer of N-acetylalanylglycine amide showing a type II beta turn. d. Conformer of N-acetylalanylglycine amide with one 1:3 ring. e. Extended N-acetylalanylglycine amide.

5.5 kcal/mole. A type I′ beta turn structure is also higher than structure a by 5.5 kcal/mole; a structure featuring a type II′ beta turn is higher than structure a by 7 kcal/mole; a type III beta turn structure is higher by 10 kcal/mole than structure a; and a nonhydrogen-bonded structure is higher by 7.5 kcal/mole. These values are approximate since the optimization of these structures was within 0.6 kcal/mole of the minimum. Indeed, as shown by Zimmerman and Sheraga (26), structures higher in energy by more than 3 kcal/mole than the most stable conformation contribute little to the partition function.

The structures a–e feature similar bond lengths and angles, with the exception of the lengthening of the N–H and C=O bonds when they are involved in hydrogen bonding. The 1:3 hydrogen bond H–O distances are about 1.7 Å, while the 1:4 hydrogen bond H–O distances are about 1.9 Å.

The fact that structure a is the lowest in energy agrees with the results of Zimmerman and Scheraga (26). Other studies of AC–Ala–NH_2 have established the stability of the C7 ring (27–28). Structure b is higher in energy than structure a by only 0.9 kcal/mole. The type II beta turn structure (structure c) is higher than structure 2 by 0.9 kcal/mole. Structures d and e are higher than structure a by more than 3 kcal/mole, with structure e not even being a true minimum. Structure 5′, which has optimized dihedral angles and is not fully extended, features the same energy as structure d, that is, 3.2 kcal/mole higher than structure a. Accordingly, structures a–c are expected to predominate in the ensemble of conformations adopted by Ac–Ala–Gly–NH_2 in gas phase or in a nonpolar solvent.

Zimmerman and Scheraga (29) showed that librational entropy tends to stabilize extended structures and destabilize turn structures. They also have calculated that Ac–Ala–Gly–$NHCH_3$ features a type II beta turn for its most stable conformer. In this structure, Ala exhibits phi and psi angles close to the standard type II beta turn values, but the glycine phi and psi resemble more a C7 ring. They describe a structure with a definite 1:4 hydrogen bond as having a slightly higher energy. In these calculations (29), the type II beta turn structure is more stable than the type I beta turn structure.

The small energy difference between the type I and type II beta turn structures of N-acetylalanylglycine amide might be compensated by solvent effects, leading to an equilibrium ratio of 1:1 for the two conformers. Indeed, NMR studies of cyclo ALA-Gly-Aca (where Aca is epsilon aminocaproic acid) (30) showed that the ratio of type I to type II depends on the nature of the solvent. In water, type I is about 35%, while in dimethylsulfoxide it is less. It is clear, though, that in both cases the type II beta turn is more stable.

This disagreement between the ab initio calculations and experiment might be due to entropy differences and to solvent effect. In order to elucidate this matter, Sapse et al. (31) have estimated both entropy and solvent effects, calculating the free energy of different conformers in solution.

In order to evaluate the solvent effect, Sapse et al. (31) used a modified Born equation. The solute molecule is placed in a cavity of the solvent of spherical or prolate shape, and the electrostatic energy of interaction is calculated. The shape of the cavity is chosen to be spherical for quasi-spherical molecules and prolate for elongated molecules. For the conformers of N-Acetylalanylglycine amide it is clear that a prolate cavity is more appropriate. The solvent is treated as a continuum and characterized by its dielectric constant. If the molecule is encapsulated in a prolate spheroidal cavity whose semimajor and semiminor axes are denoted by a and b, respectively, f was defined as $f = (a^2 + b^2)\frac{1}{2}$, and $\xi_0 = a/f$. In this case the interaction energy takes the form (32)

$$U = \frac{1}{2}\left[\frac{1}{\varepsilon} - 1\right]\sum_{n=0}^{\infty}\sum_{m=0}^{\infty}\frac{F_n^m}{\Delta_n^m}\left[(\gamma_n^m)^2 + (\sigma_n^m)^2\right],$$

where the symbols are

$$\gamma_n^m = \sum_{j=1}^{N} Q_j P_n^m(\xi_j)P_n^m(\eta_j)\cos(m\phi_j),$$

$$\sigma_n^m = \sum_{j=1}^{N} Q_j P_n^m(\xi_j)P_n^m(\eta_j)\sin(m\phi_j),$$

$$F_n^m = \frac{1}{f}[2 - \delta_{m0}](-1)^m(2n + 1)\left[\frac{(n - m)!}{(n + m)!}\right]^2,$$

$$\Delta_n^m = \frac{P_n^m(\xi_0)}{Q_n^m(\xi_0)} - \frac{1}{\varepsilon}\frac{[P_n^m(\xi_0)]'}{[Q_n^m(\xi_0)]'}.$$

In the above equation ξ_j, η_j, and ϕ_j are the prolate spheroidal coordinates of the jth ion, and P_n^m and Q_n^m are associated Legendre functions of the first and second kind, respectively.

The net atomic charges and the position of the atoms were determined by ab initio calculations, using Mulliken population analysis for the former and geometry optimization for the latter. This study uses in this respect the results of (23).

The radius of the cavity was obtained in the following way: The two farthest atoms were determined. Then the center of the cavity was placed midway between them. This point becomes the origin of the coordinate system. Coordinate axes are rotated in such a way that the new z-axis is along the line joining the two farthest atoms. Then the smallest rectangular box that can contain the molecule is determined. The most highly charged atoms in contact with the solvent are the oxygen atoms. To ensure a close contact between the oxygens and the solvent, the origin is shifted to the center of the three oxygens. Finally, the semimajor and semiminor axes of the prolate spheroid are obtained by a Monte Carlo method that makes the area of the surface of the cavity a minimum. A Van der Waals radius of 1.3 A is added to the values of the semiaxes. This value is intermediate between the Van der Waals radius of hydrogen and that of oxygen.

The entropies were calculated using Scheraga's method (26), where the derivatives of the energy with the dihedral angles was obtained from ab initio calculations (23). The calculations indicate that structures b and c (nomenclature of (23)) are almost equal in electrostatic interaction with the solvent, which in this case is water. This is to be expected, since they have about the same shape, and in both, out of three oxygens, two are free to be in contact with the solvent, and the third, even though it is hydrogen-bonded, is quite close to the solvent. Structure a has two oxygens that are hydrogen-bonded and quite protected from the solvent. Structure d has an oxygen hydrogen-bonded in a way that hides it from the solvent, and another one protected by a CH_3 group. In general, the solute–solvent interactions are somewhat underestimated, since the STO-3G basis set underestimates the net atomic charges.

The entropy calculations show the highest librational entropy for structure c, where the hydrogen bond is set at some distance from the rest of the molecule. Structure b has a lower entropy due to more crowding, while structure 4, which features a C7 single ring, has an even smaller entropy. As expected, structure a, which features a double C7 ring, is the lowest in entropy.

The sum of these effects leads to the enhanced stability of structure c, which features a type II beta turn, in agreement with the experimental data (30).

In the early 1980s, Schafer and coworkers began to perform ab initio calculations on model dipeptides, such as N-acetyl N′-methyl derivatives of glycine and alanine (33,34,35). These calculations used the 4-21G basis set, at Hartree–Fock level. As pointed out by Jiang et al. (36), one of the most important quantitative results obtained by geometry optimization of peptides is related to the determination of the phi and psi angles, since the bond lengths and bond angles in peptides depend on the torsional angles, especially on the flexibility of phi amd psi. These angles change their size significantly when the peptide undergoes a change, for instance from a type I beta turn to a type II beta turn. However, even bond angles such as the alpha angle, N-Calpha-C′, undergo changes in these cases.

Schafer et al. (37) discovered that the conformational geometry maps calculated for Ala can be used as a basis for the prediction of backbone structural parameters of proteins. These predictions turned out to be as accurate as the high-resolution protein crystallography data (36). When the alpha angle is concerned, Schafer et al. (37) found that the root mean square deviation between experimental and theoretical data was 1.67° for crambin (38) and 1.6–2.1° for deaminooxytocin (39).

Jiang et al. (36) calculated the alpha angle of about 40 proteins for which the crystal stuctures were available, in order to estimate the phi/psi correlation of the deviation from ideal geometry and to compare it with ab initio results.

They used the 4-21G basis set at Hartree–Fock level to optimize the geometry of 144 different structures of Ala, using a phi/psi grid with increments of 30°. For each fixed value of phi and psi, the other parameters of the molecule were optimized (36). The optimized coordinates of Ala were subsequently used to derive analytical functions for the most important bond lengths and angles, such as $N-C_\alpha$, $C_\alpha-C'$, and $N-C_\alpha-C'$. These funtions were expanded in terms of natural cubic spline parameters, incorporated in a program for peptide simulation (PEPSI) that permits the calculation of bonds lengths and angles at any point on the phi/psi surface (36).

Jiang et al. (36) used for comparison protein crystal structures taken from the Brookhaven data bank. The lowest-temperature conformations were chosen, and in few cases mutants were included. The spreadsheet function of the Biosym Discover/Insight II modeling program was used in order to obtain the set of parameters ψ_i, ϕ_i and $(N-C_\alpha^- C') = \alpha_i$. The set of ϕ and ψ were put in the PEPSI program, and α_i was calculated. These results were compared with the α_i values from the crystallographic data, and the root

mean square deviations $\delta = \left[\dfrac{1}{m} \sum_i^m (\alpha_i^{cryst} - \alpha_i^{cale})^2 \right]^{1/2}$ were calculated for each

protein. Here m is the number of residues in a given protein.

As shown by Jiang et al. (36), this algorithm has some problems:

- The calculated values were derived from a model dipeptide (Ala), while the comparison was made with residues from an extended peptide.
- Ab intio calculations do not take into account the vibrational energy, while the proteins include the zero-point vibrational energy and are vibrationally averaged at some temperature that is not absolute zero.
- The calculations were performed for alanine, while the protein contains many other residues.

In spite of these problems, the authors of the study find close agreement between experimental and calculated parameters. The best agreement is found for crambin, which features a δ of 1.67°.

The discrepancies for the alpha-helical and beta regions are different: In the alpha regions the calculated angles are 1.5° above the crystal angles, while in the beta regions they are 2.1° below. This kind of systematic deviation (36) might be due to cooperative effects in extended chains.

The examination of the results, as concluded by Jiang et al. (36), demonstrate that there is a correlation between the ψ, ϕ torsion angles of proteins and the extension of their alpha angles. The crystallographic phi/psi correlations are in good agreement with the ab initio predictions.

The authors of the study stress the importance of an accurate parametrization of the conformational energy trends. Without correct bond lengths and angles, interactions cannot be properly described, and model properties are uncertain (36). The ability of ab initio methods to predict structural parameters as functions of the torsion angles is an important development for computational chemistry (36).

The study of Jiang et al. (36) stresses the need for phi/psi correlated values and for a library of detailed functions of this type. Such a library could explain serious discrepancies between calculated parameters and experimental ones, by helping with the interpretation of the experimental results (36). Jiang et al.'s (36) study might prove to be the beginning of a new type of study on protein structures.

In a subsequent study, Jiang et al. (40) compare the crystallographic $N-C_\alpha$, $C_\alpha-C'$, and $N-C_\alpha-C'$ backbone parameters of 43 oligopeptides and the $N-C_\alpha-C'$ angles of 37 proteins to the corresponding parameters in N-acetyl N'-methyl alanine amide, as determined by ab initio calculations.

Schafer et al. (41) point out that accurate standard geometries, as determined by ab initio methods, should consist of standard geometry functions. Indeed, molecular geometry is local, that is, it depends on where the molecule is in conformational space (40). It has been shown (33) that conformational geometry maps are as important as conformational energy maps for peptides.

As pointed out by Jiang et al. (40), the ab initio method includes a series of artificial circumstances, such as describing isolated, vibrationless systems in vacuum. Therefore, the comparison of calculated data with experimental results requires corrections. When the experiments are performed in vapor phase, it is to be expected that the calculations will agree more with the experiments. However, even in crystalline phase it was found that some parameters such as the $N-C_\alpha-C'$ angle, as described before, take similar values as calculated by ab initio (HF/4-21G) calculations and X-ray crystallography (36) in a large number of proteins.

Jiang et al. (40) calculated the $N-C_\alpha$ and $C_\alpha-C'$ bond lengths and the $N-C_\alpha-C'$ bond angle for a series of 43 oligopeptides. They compared these data with the corresponding parameters found in crystal phase and deposited in the Cambridge Crystallographic Data File (CCDF) (42). In this study all nonnatural amino acid residues were excluded, as well as glycine and proline, which cannot be compared to alanine.

The ab initio work uses the HF/4-21G method and models the dependence of the above-mentioned parameters on the ϕ and ψ torsional angles. In this way, conformational geometry maps are constructed at points situated on a grid with $30°$ increments in the ϕ/ψ space (40). Thus 144 structures of Ala are obtained, and from them analytical functions were derived for $N-C_\alpha$, $C_\alpha-C'$, and $N-C_\alpha-C'$. These functions were expanded in terms of natural cubic spline parameters and incorporated into the PEPSI program, which can be used to calculate the values of the three parameters at any point on the ϕ/ψ surface.

Although the calculations were performed on the single model dipeptide Ala, even though the crystal structures belonged to extended peptide sequences containing residues of different types, the root-mean-square deviation between the calculated and experimental geometrical parameters is relatively small. The authors (40) conclude that the torsion-dependent $N-C_\alpha-C'$ angle functions are more appropriate for various areas of protein study, such as empirical modeling procedures, rather than standard constants.

A study by Karplus (43) in which protein geometry is investigated as a function of the ϕ/ψ torsional angles confirms the above results.

Indeed, the calculated geometry surfaces are useful as tools for the identification of torsional trends in experimental parameters. They establish mean values around which the values of the parameters can be found, although there might be large differences in individual proteins (40).

References

1. Beachy, M.D., Chasman, D., Murphy, R.B., Halgren, T.A., and Friesner, R.A. *J. Am. Chem. Soc. 119*, 5908, 1997.
2. Mohamadi, F., Richards, N.G.J., Guida, W.C., Liskamp, R., Liptom, M., Caulfield, C., Chang, G., Hendrickson, T., and Still, W.C. *J. Comp. Chem. 11*, 440, 1990.
3. Dunning, T.H. *J. Chem. Phys. 90*, 1007, 1989.
4. Murphy, R.B., Beachy, M.D., Friesner, R.A., and Rignalda, M.N. *J. Chem. Phys. 103*, 1481, 1995.
5. St.-Amant, A., Corne, W.D., Kollman, P.A., and Halgren, T.A. *J. Comp. Chem. 16*, 1483, 1995.
6. Murphy, R.B., Pollard, W.T., and Friesner, R.A. *J. Chem. Phys. 106*, 5073, 1997.
7. Gould, I.R., Cornell, W.D., and Hillier, I.H. *J. Am. Chem. Soc. 116*, 9250, 1994.
8. Head-Gordon, T., Head-Gordon, M., Frisch, M.J., Brooks, C.L. III, and Pople, J.A. *J. Am. Chem. Soc. 113*, 5989, 1991.
9. Frey, R.F., Coffin, J., Newton, S.Q., Ramek, M., Cheng, V.K.W., Momany, F.A., and Schafer, L. *J. Am. Chem. Soc. 114*, 5369, 1992.
10. Moult, J., and James, M.N.G. *Proteins 1*, 125, 1986.
11. Jorgensen, W.L., Laird, E.R., Nguyen, T.B., and Tirado-Rives, J. *J. Comp, Chem. 14*, 206, 1993.
12. Halgren, T.A. *J. Comp. Chem. 17*, 490, 520, 553, 1996.
13. Halgren, T.A., and Nachbar, R.B. *J. Comp. Chem. 17*, 587, 1996.
14. Boys, S.F., and Bernardi, F. *Mol. Phys. 19*, 553, 1970.
15. Dreyfus, M., Maigret, B., and Pullman, A. *Theor, Chim. Acta. 17*, 109, 1970.
16. Schlegel, H.B., Gund, P., and Fluder, E.M. *J. Am. Chem. Soc. 104*, 53, 1982.
17. Zielinski, T.J., and Poirer, R.A. *J. Comp. Chem. 5*, 466, 1984.
18. Johansson, A., Kollman, P., Rothenberg, S., and McKelvie, J. *J. Am. Chem. Soc. 96*, 3794, 1974.
19. Scheiner, S., and Kern, C.W. *J. Am. Chem. Soc. 99*, 7042, 1977.
20. Hinton, J.F., and Harpool, R.D. *J. Am. Chem. Soc. 99*, 349, 1977.
21. Sapse, A.M., Fugler, L.M., and Cowburn, D. *Int. J. Quant. Chem. 29*, 1241, 1986.
22. Schulz, G.E., and Schirmer, R.H. *Principles of Protein Structure*, Springer-Verlag, New York, 1979.
23. Sapse, A.M., Daniels, S.B., and Erickson, B.W. *Tetrahedron 44*, 999, 1988.
24. Erickson, B.W., Daniels, S.B., Reddy, P.A., Unson, C.G., Richardson, J.S., and Richardson, D.C. In *Computer Graphics and Molecular Modeling*, Zoller, M., and Flettering, R., (eds.) Cold Spring Harbor Laboratory, Cold Spring Harbor, NY, 1986.

25. Venkatachalam, C.M. *Biopolymers 6*, 1425, 1968.
26. Zimmerman, S.S., and Scheraga, H.A. *Biopolymers 16*, 811, 1977.
27. Schafer, L., Klimkowski, V.J., Momany, F.A., and Chuman, H. *Biopolymers 23*, 2335, 1984.
28. Beveridge, D.L., Ravishanker, G., Mezei, M., and Gedulin, B. In *Biomolecular Stereodynamics 3*, Sarma, R.H., and Sarma, M.H., (eds.) Adenine Press, Guiderland, NY, 1986.
29. Zimmerman, S.S., and Scheraga, H.A. *Biopolymers 17*, 1849, 1978.
30. Deslauriers, R., Evans, D.J., Leach, S.J., Meinwald, Y.C., Minasian, E., Nemethy, G., Rae, I.D., Scheraga, H.A., Somorjai, R.L., Stimson, E.R., Van Nispen, J.W., and Woody, R.W. *Macromolecules 14*, 985, 985, 1981.
31. Sapse, A.M., Jain, D.C., deGale, D., and Wu, T.C. *J. Comp. Chem. 11*, 573, 1990.
32. Gersten, J., and Sapse, A.M. *J. Am. Chem. Soc. 107*, 3786, 1985.
33. Schafer, L., Van Alsenoy, C., and Scarsdale, J.N. *J. Chem. Phys. 76*, 1439, 1982.
34. Scarsdale, J.N., Van Alsenoy, C., Klimkowski, V.J., Schafer, L., and Momany, F.A. *J. Am. Chem. Soc. 105*, 3438, 1982.
35. Schafer, L., Klimkowski, V.J., Momany, F.A., Chuman, H., and Van Alsenoy, C. *Biopolymers 23*, 2335, 1984.
36. Jiang, X., Cao, C., Teppen, B., Newton, S.Q., and Schafer, L. *J. Phys. Chem. 99*, 10521, 1995.
37. Schafer, L., Cao, M., and Meadows, M.J. *Biopolymers 35*, 603, 1995.
38. Teeter, M.M., Roe, S.M., and Heo, N.H. *J. Mol. Biol. 230*, 292, 1993.
39. Wood, S.P. et al. *Science 232*, 633, 1986.
40. Jiang, X., Yu, C.H., Cao, M., Newton, S.Q., Paulus, E.F., and Schafer, L. *J. Mol. Structure 403*, 83, 1997.
41. Schafer, L., Ewbank, J.D., Klimkowski, V.J., Siam, K., and Van Alsenoy, C. *J. Mol. Structure 135*, 141, 1986.
42. Allen, F.H., Davies, J.E., Galloy, J.J., Johnson, O., Kennard, O., Macrae, C.F., Mitchell, E.M., Smith, J.M., and Watson, D.W. *J. Chem. Info. Comp. Sci. 31*, 187, 1991.
43. Karplus, M. *Prot. Sci. 5*, 1406, 1996.

12
Oligopeptides That Are Anticancer Drugs

The peptidic bond is present in some molecules that have been found to possess anticancer activity. These oligopeptides have been termed "lexitropsins" or information-reading molecules. They are also antibiotics, and evidence from biochemical pharmacology indicates that they act to block the template function of DNA. They do so by binding selectively to adenine-thymine (AT) sequences in the minor groove (1).

The regulation of gene expression by control proteins, such as promoters and repressors, in prokaryotes as well as in eukaryotes, requires the specific recognition of both single-stranded and double-stranded nucleic base sequences (2–4). Control proteins and xenobiotics appear to utilize two different channels of information in reading the base sequences in double-helical DNA. Although control proteins use in general the major groove, certain polymerases and antibiotics bind to the minor groove. The reason for the selection of the major groove by control proteins is probably the fact that the number of hydrogen bonds and their discriminatory capacity is higher; therefore, higher information content is available. However, the minor groove offers advantages in the sense that it is less occupied than the major groove and thus more accessible to attack. This might be the reason for the evolutionary development of minor-groove-selective antibiotics by microorganisms to combat competitors.

Examples of naturally occurring minor-groove-binding antibiotics include netropsin (5), distamycin (6), anthelvencin (7), and kikumycin (8). Analysis of their structural requirements for DNA recognition suggest that appropriate modification in their structure can alter the DNA sequence recognition. As such, compounds have been synthesized that bear hydrogen-bond-accepting heterocycles, among which are imidazole, furan, 1,2,4 triazole, and thiazole. Using NMR analysis, complementary strand footprinting, and additional spectroscopic techniques (9–13), it was found that these compounds exhibit somewhat altered sequence recognition.

Distamycin and netropsin bind in the minor groove of DNA with high specificity for A-T sequences. When they are attached to cleaving agents (for example, EDTA-Fe 11), they form affinity cleaving agents with A-T specificity (14). Both molecules are active against a number of tumors. However, they cannot be used in chemotherapy because of their high toxicity.

Netropsin and distamycin are oligopeptides built from 4-amino-1-methylpyrrole-2-carboxylic acid residues and strongly basic side chains. They differ in that netropsin has two pyrrole residues and distamycin has three, and while netropsin ends on one side with a guanidinoacetyl side chain and in the other with an amidinopropyl, distamycin has only the aminopropyl side chain. As determined by X-ray studies, the molecules exist in a curved shape, with the N-methyl on the convex side and the amide NH groups on the concave side. The torsion angle between the pyrroles at the peptidic bond is about 20° (15–16). At physiological pH, distamycin features one positive charge and netropsin features two positive charges.

Numerous experimental studies have investigated the way netropsin and distamycin bind to DNA and various synthetic polynucleotides. The methods used include UV absoption, thermal melting, hydrodynamics measurements, and Cotton effects in ORD and CD (17). The main methods of investigation are NMR and X-ray diffraction analysis of the complexes formed by DNA and the minor groove binder. The DNA used for binding is B-DNA, and all the methods mentioned concur in the finding that A-T sequences are preferred. When experiments were performed with A and Z DNA, it was found that they transform into B-DNA upon complexation (18–19). It has been found that there is no gross helical distortion of the DNA fragments on binding.

Goodsell and Dickerson (20) have discussed the reasons for the preference of the lexitropsins and their parent molecules for A-T sequences. As shown by X-ray analysis, the drug molecule sits with the two pyrrole rings in the minor groove, with the CH pointing down the floor of the groove. They are packed close to the adenine hydrogen atoms, and if the adenine were substituted by a guanine, there would be strong steric hindrance with the NH_2 group of guanine. Indeed, the components of molecular recognition that determine the sequence selectivity of the drugs include van der Waals nonbonded contacts between the ligand and the surface of the minor groove. In addition, the binding includes electrostatic attraction between the ligand and DNA and hydrogen-bond formation between the amide NH groups that bridge the rings to the exposed adenine N_3 and thymine O_2 on adjacent sites and between inward-directed heteroatoms and different sites on the DNA.

Pullman and coworkers (21–22) indicate that the binding of the drug to the DNA fragment is due not only to hydrogen-bond formation but especially to electrostatic interactions between the positive charges on the drug and the negative potential in the minor groove of DNA. This fact indicates a supplementary reason for the A-T preference of the ligand, besides the steric hindrance introduced by the NH_2 group of guanine, namely, the fact that minor grooves with A-T content feature a higher negative electrostatic potential than the G-C minor grooves. However, evidence for the presence of hydrogen bonds has been provided by Raman spectroscopy (24).

When netropsin binds to the minor groove, it replaces the water molecules found otherwise in the groove. Changes in the double helix consist of slight widenings of the minor groove and a slight bend (of about 8°), but no unwinding or elongation has been observed (17).

The binding of netropsin to a CGCGAATTCGCG DNA fragment occurs with the formation of complexes that are similar in solution and in crystalline state (17). As far as distamycin is concerned, its binding to the same DNA fragment in crystal phase is similar to that of netropsin, except that the direction of the ligand in the groove is reversed. In solution, though, a ratio of 1:1 ligand to DNA and 2:1 ligand to DNA coexisted (17).

When a DNA fragment with the sequence CGCGATACGCG was used, X-ray analysis showed that netropsin binds to the minor groove equally well in both orientations, in agreement with MR results (25).

To investigate the possibility of changing the sequence preference, the pyrrole rings were replaced by imidazole rings, otherwise retaining the structure of netropsin (26). It was found indeed that there is better GC binding, which was attributed to the possibility of formation of hydrogen bonds between the imidazole nitrogens and the NH_2 group of guanine.

In order to apply the netropsin and distamycin drugs clinically, attempts were made to modify them in a way that would decrease their toxicity. Conjugation with intercalators, formation of bis-netropsins, and substitutions by lipophilic groups were tried (27).

One of the methods applied to the study of lexitropsins is theoretical investigation. Molecular-mechanics studies were performed by Pullman et al. (22). Ab initio methods were also used for a series of compounds.

Two of the small lexitropsins investigated with ab initio calculations were amidinomycin and noformycin (28). Amidinomycin was isolated from *Streptomyces flavochromogenes* and was shown to present the 1R,3S configuration. Noformycin, which features only one chiral center, was isolated from *Nocardia formica* and shown to feature the 4S(+) configuration. Both have been shown to exhibit activity against a variety of plant and animal viruses. These two molecules feature an amidine group at one end. Noformycin features an aminopyrolidinium ion at the other end, and amidinomycin features an aminocyclopentene ring. Both molecules feature one peptidic bond, between the rings and the rest of the molecule. Larger molecules feature greater biological potency, but shorter oligopeptides seem to be more discriminating in terms of sequence preference.

A point of great interest in these molecules, besides their size, is the presence of chiral carbons, which may play an important role in DNA recognition. Indeed, chiral ligands should be isohelical with the minor groove, and as such, marked differences were observed in the binding of 4S(+) anthelvencin and dihydrokikumycin, which are natural products, and their 4R enantiomers. As the size of the molecule decreases and the chiral center becomes a larger part of the molecule, the difference in binding increases.

The electric charges on the molecules are important for binding. When the amidine group accepts a proton and becomes a positive amidinium ion, noformycin and amidinomycin become bis-cationic and bind strongly to DNA featuring A-T sequences. Amidinomycin and noformycin have large pKa values at both basic sites and are bis-cationic at physiological pH.

Sapse et al. (28) have subjected noformycin and amidinomycin to complete geometry optimization, using ab initio calculations at Hartree–Fock level with the 6-31G basis set. The energies and the net atomic charges were also calculated. The geometrical parameters obtained for amidinomycin were in good agreement with experimental data. No experimental data were found for noformycin. In both molecules the rings exhibit puckering, with the cyclopentane ring of amidinomycin lying orthogonal to the plane of the rest of the molecule.

The molecules in their most stable configuration were displayed via molecular modeling using the Quanta computer graphics program and were inserted into the minor groove of a DNA fragment. This fragment was GCGAATTCGC from which the water molecules were removed. The binding energy is the sum of the van der Waals and electrostatic interactions.

For amidinomycin, the 1S-3R, 1R-3R, 1R-3S, and 1S-3S isomers were considered. The isomer that featured the strongest binding was the 1S-3R isomer. The next in order of strength of binding, quite similar to the first, was RR, followed by RS and SS. It appears from the results that the second chiral center is the most responsible for the difference between the binding of various isomers.

For noformycin there is a larger difference between the binding of S and R isomers, with the S isomer binding more than the R one. In general, noformycin exhibited more binding than amidinomycin. This is probably due to stronger hydrogen-bond formation and to a greater dipole moment. Indeed, for noformycin the ab initio calculations predict a dipole moment of 7.69 debyes, while for amidinomycin the predicted dipole moment is 5.58 debyes. Both molecules show preference for the AT sequences but also bind to GC sequences.

One of the most important factors determining effective minor-groove binding of ligands is the repeat distance of the nucleotide units of DNA and the van der Waals and hydrogen-bond contacts generated by the interactions. These are referred to sometimes as the phasing problem. As the ligands increase in size, the hydrogen bonds and the van der Waals contacts become out of phase with the spacing between nucleotides. In the case of noformycin and amidinomycin (for the 1S,3R isomer), the N–N distances are close to the ideal value found for a potent minor-groove binder. In contrast, the natural isomer of amidinomycin, which is 1R,3S, features too large N–N distances, and this might explain the reduced binding.

Ab initio calculations have also been applied to certain prototype lex-itropsins in which one of the N-pyrrole moieties of netropsin was replaced by imidazole (23). For this molecule, too large to be treated at ab initio level in its entirety with a double-zeta basis set, the following procedure was applied: The molecule was split into three overlapping segments that were geometry-optimized at Hartree–Fock level using the 6-31G basis set. The intersegment parameters, including the torsion angles, were optimized using the STO-3G basis set with the rest of the parameters at their 6-31G optimum values. The conformation with an angle of 63° between the central rings, with the guanidinium group coplanar with the ring it is attached to and with the amidinium group set at 99° to the ring it is attached to, was found to be the most stable.

The optimized molecule was modeled with the Quanta computer graphics program and inserted into the minor groove of a B-DNA fragment. This fragment is GCGAATTCGC and its corresponding strand. The water molecules from the minor groove of the fragment were removed. The insertion was performed via the use of the rotational–translational matrix, aiming for a minimum in the energy of the complex. Besides the optimized conformation, two other conformations were docked into the same type of B-DNA fragments. One of them was a completely planar conformer, and the other kept the torsion angle between rings at 63° but set the the amidinium group coplanar with the ring to which it was attached. These two conformers were higher in energy than the most stable conformer by 4.5 kcal/mole and 7.9 kcal/mole, respectively. The last con-former exhibited the best binding to DNA because the guanidinium and amidinium groups did not engage in steric clashes with the interior of the minor groove.

The molecule binds somewhat to DNA fragments containing GC sequences. However, in spite of the additional hydrogen-bond formation possibility, the binding is substantially lower (23).

The concept of replacing the pyrrole rings that occur in natural com-pounds by other heterorings raises the question of how willing the het-erorings are to accept hydrogen bonding. This property is related to their proton affinity. Kabir and Sapse (29) applied ab initio calculations to the determination of the proton affinities of imidazole, oxazole, and thiazole. In addition, N-methyl imidazole and imidazole featuring a formamyl group on a carbon were also investigated.

Figures 12.1–12.8 display the molecules investigated. The odd-numbered figures are the neutral species, and the even-numbered are the protonated species.

The 6-31G basis set was used to perform Hartree–Fock calculations with complete geometry optimization. The proton affinities were calculated by subtracting the energy of the neutral species from that of the protonated species. Single-point calculations at the 6-31G optimized geometries were performed with the 6-31G* basis set.

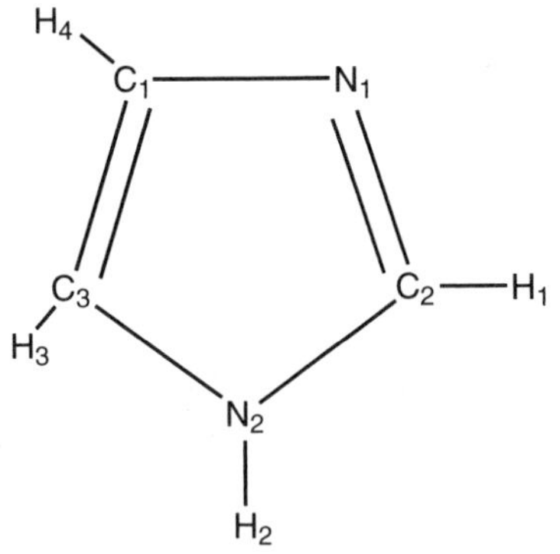

FIGURE 12.1 Imidazole.

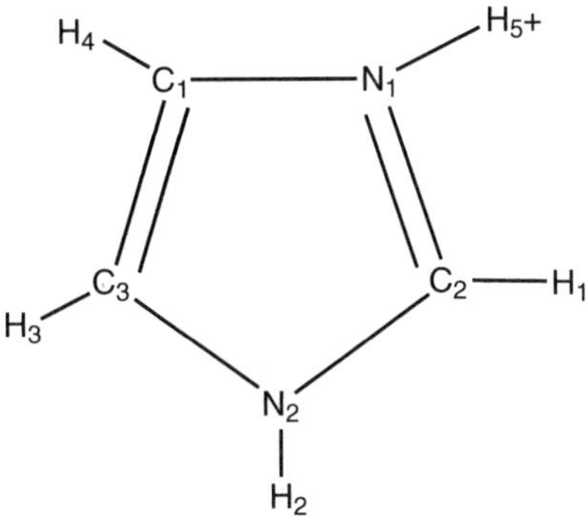

FIGURE 12.2 Protonated imidazole.

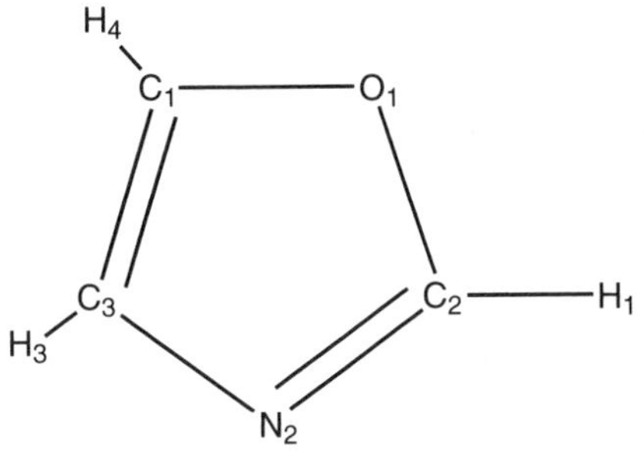

FIGURE 12.3 Oxazole.

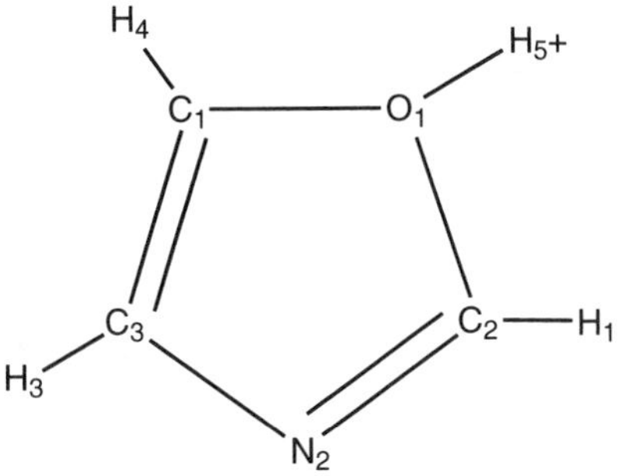

FIGURE 12.4 Protonated oxazole.

The proton affinities as obtained with both basis sets decrease substantially from imidazole to oxazole to thiazole. Indeed, although the imidazole molecule features a proton affinity of 245.92 kcal/mole at 6-31G level and 239.45 kcal/mole at 6-31G* level, the oxazole features 172.87 kcal/mole and 164.47 kcal/mole at 6-31G and 6-31G* levels, respectively, and the thiazole has corresponding values of 137.11 and 144.45 kcal/mole. It is thus clear that nitrogen-containing rings will be more prone to proton acceptance and so to hydrogen bond formation than the other rings. It has indeed been shown that lexitropsins that contain thiazole rings particularly avoid binding to GC sequences because of the large size of the sulfur ion, which increases the steric hindrance.

The presence of a methyl group on one of the nitrogens of the imidazole ring increases somewhat the proton affinity of the other nitrogen. Conversely, a peptide substituent on the carbon between the two nitrogens decreases the proton affinity of the ring (29).

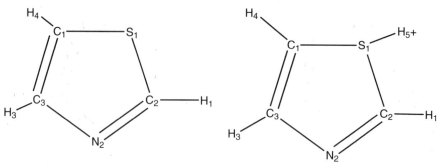

FIGURE 12.5 Thiazole.

FIGURE 12.6 Protonated thiazole.

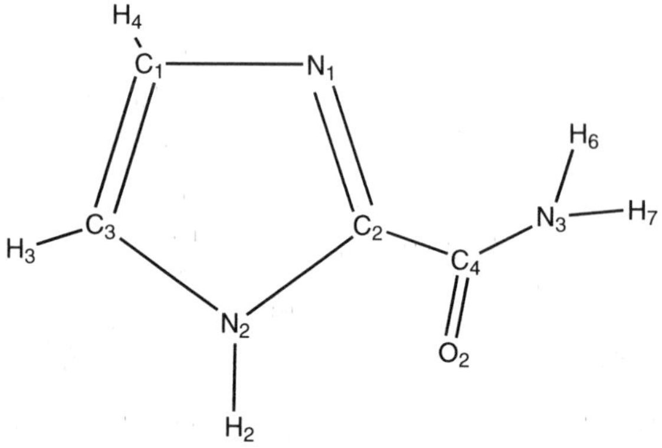

FIGURE 12.7 Amideimidazole.

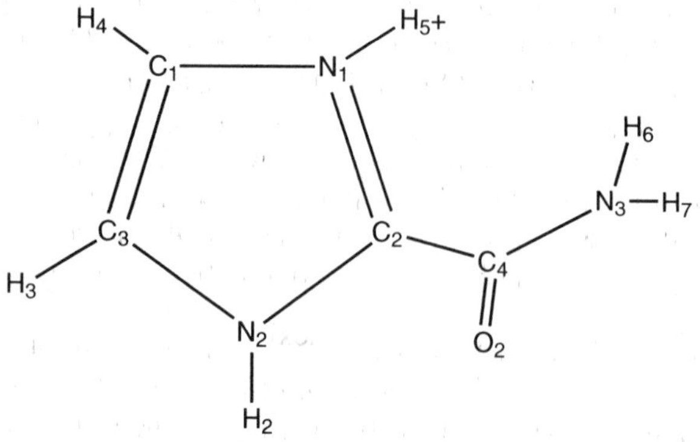

FIGURE 12.8 Protonated amide imidazole.

One of the useful features of the anticancer oligopeptides is their ready uptake into cells and concentration in the cell nucleus (30). However, a limitation of the applications of these molecules is their susceptibility to degradation by peptidase (31). To avoid degradation, the attention of the researchers has been focused on the development of amide bond surrogates in order to be able to produce compounds resistant to degradation by enzymes.

Since the introduction of the Lawesson reagent (32), by which amides are converted into thioamides, a number of thiopeptide analogues of biological substrates and regulators have been investigated. Indeed, the thioamide function is resistant to enzymatic attack, as seen for the case of thioamide analogues of caboxypeptidase A (33). The structure of distamycin/DNA complex, as determined by X-ray studies, has proved that the carbonyl groups are away from the floor of the groove; therefore, replacing of oxygen by sulfur should not affect the binding (34). Zimmerman et al. (30) synthesized thioformyldistamycin and reported a study of its resistance at the thioamide group to enzymatic hydrolysis. NMR results indicate that in solution there are two forms of the drug: E and Z conformers about the thioamide bond (30). Using complementary strand footprint methodology, they examined the DNA sequence selectivity of the two conformers. In addition, they studied acid catalyzed hydrolysis.

Zimmerman et al. (30) found that thioformyldistamycin as well as distamycin are susceptible to acid and peptidase catalyzed hydrolysis. The amide bond in the latter, and to a lesser extent in the former, can be hydrolyzed. The formyl amide bond in distamycin is also susceptible to hydrolysis, and the thioformyl surrogate is unreactive. Both compounds have been found to bind to DNA with comparable strength and affinity (30).

Since the NMR data indicate the presence of the E and Z conformers, it was deemed interesting to determine the preferred conformation of the thioformyl group. To this purpose, a rotational search was carried out around the N–C bond (30). The method of the search was ab initio at Hartree–Fock level using the 3-21G* basis set. The 3-thioformylamino-N-Methylpyrrole was used as a model for the entire molecule. The E and Z conformations were used as starting geometries, and all the parameters of the molecule were allowed to relax. The E conformer was found to be more stable than the Z conformer by 17.3 kcal/mole. The barrier to interconversion between the two conformers was estimated by setting the SCNC torsion angle (which is 180° for E and 0° for Z) to a 90° value. This structure proved to be higher in energy than E by 26.3 kcal/mole.

If it is accepted and understood that lexitropsins and their parent molecules prefer A-T sequences for binding, it has also been found that there are differences between the specific A-T sequences to which various oligopeptides bind. For example, footprinting and affinity-cleaving studies (14) have shown that distamycin prefers sites that contain multiple adjacent adenines, while NMR and calorimetric studies of netropsin and distamycin A indicate a strong binding at AATT sites (35).

One important aspect of the binding of these compounds to DNA is the presence of positive groups at one or both ends of the long flexible strands of the ligand. Anthelvencin, kikumycin, and, as mentioned before, other small lexitropsins feature an aminopyrrolidinium group at one end and an amidinium at the other. Netropsin features a guanidinium group at one end and an amidinium group at the other. These groups provide a positive potential that binds electrostatically to the negative potential present in the minor groove of DNA. Moreover, they might form hydrogen bonds with the DNA bases, contributing thus to the binding of the whole molecule. There is also the possibility that they act as intercalators. Differences between the binding ability of these groups to the DNA bases might account for differences in the binding of the various oligopeptides. For instance, netropsin and anthelvencin are very similar, except for the nature of the group at one end. However, as shown by Lee et al. (13), these molecules show a large difference in their DNA binding, with netropsin featuring a binding energy of 53 kcal/mole and anthelvencin an energy of only 33 kcal/mole. It was considered interesting to assess how much of this difference is due to the difference in binding to the DNA bases of the two groups. Sapse et al. (36) have applied ab initio calculations to the study of this problem.

Several complexes were examined: the complex formed by thymine with the guanidinium ion through binding at the O2 of thymine, called 1a, the complex of thymine with the guanidinium ion bound at O1, called 2a, the complex formed by thymine with the aminopyrrolidinium ion, at O2, called 1b, and the same complex bound at O1, called 2b.

The calculations were performed at Hartree–Fock level using the 6-31G basis set for the optimization of the subsystems. The 6-31G geometries of the subsytems were kept frozen during the optimization of the complexes in which the O–H, the C–O–H, and the N–H–O angles were optimized with the 3-21G basis set. The dihedral H–O–C–N and N–H–O–C angles were also optimized, with starting values of $0°$, $90°$, and $180°$. The binding energy was defined as the energy of the complex minus the sum of the energies of the subsystems.

The hydrogen bond in 1a was found to deviate from linearity with an N–H–O angle of $141.8°$ instead of $180°$. This permits the formation of an additional hydrogen bond. The complexes formed by the hydrogen bond at the oxygen attached to the carbon between the two nitrogens ("a" complexes) are slightly less stable than the "b" complexes. These complexes are shown in Figure 12.9.

It was found (36) that the guanidinium binds by 2.9 kcal/mole more than the aminopyrrolidinium ion. This result indicates that the difference between the binding of lexitropsins that feature guanidinium at one end and those containing the aminopyrrolidinium ion might be due partially to

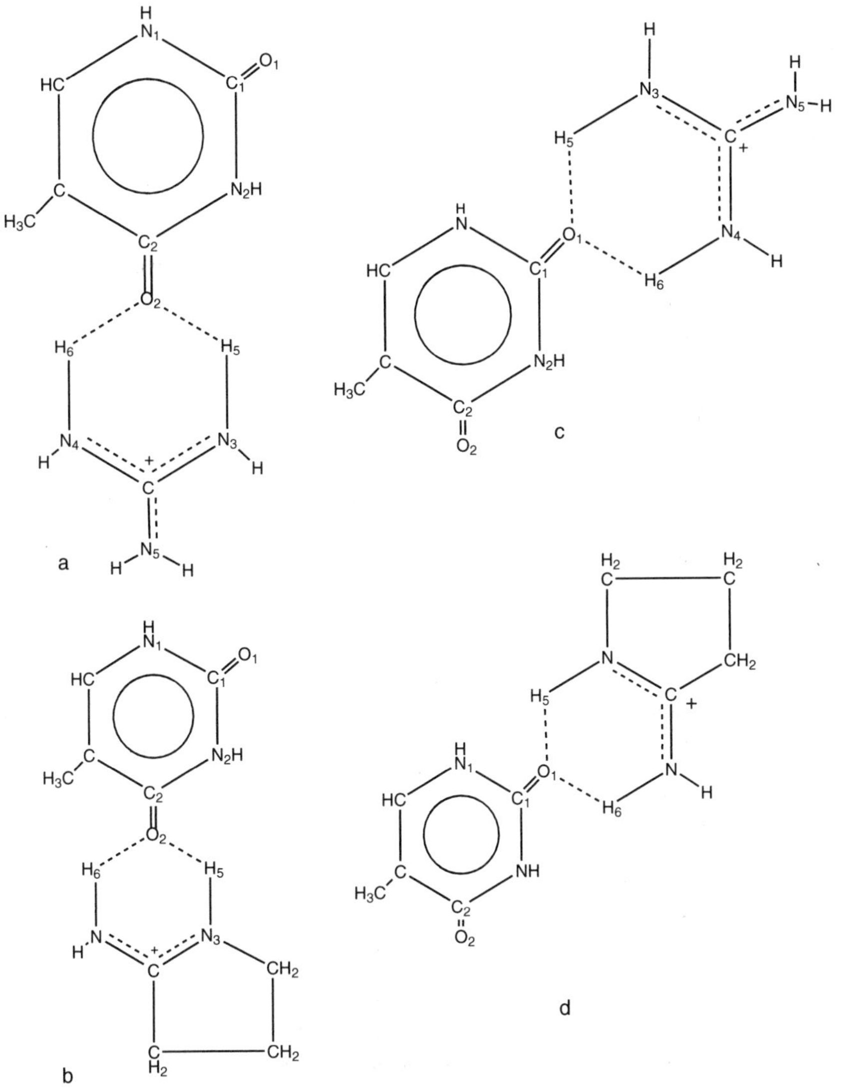

FIGURE 12.9 a. Complex formed by thymine with the guanidinium ion. b. Complex formed by thymine with the aminopyrrolidinium ion. c. Complex formed by thymine with the guanidinium ion with binding at O_1. d. Complex formed by thymine with aminopyrrolidinium ion with binding at O_1.

the difference in binding of the two groups to the DNA bases. However, the difference in binding energy of anthelvencin and netropsin is too large to be entirely justified this way.

It might be concluded that ab initio calculations can provide additional information about the structure of lexitropsins and their parent molecules as well as about their binding to DNA. Such information can eventually be used for the design of drugs with chosen sequence specificity.

References

1. Hahn, F.E. In *Antibiotics III. Mechanism of Action of Antimicrobial and Anti-tumor Agents.* Corcoran, J.W., and Hahn, F.E., (eds.) Springer-Verlag, N.Y., 1975.
2. Frederick, C.A., Grable, J., and Melia, M. *Nature 309*, 327, 1984.
3. Bishop, J.M. *Cell 42*, 23, 1985.
4. Suck, D., and Oefiner, C. *Nature 321*, 620, 1986.
5. Julia, M., and Preau-Joseph, N. *Compt. Rend. des Séances de L'Académie de Sciences 275*, 1115, 1963.
6. Arcamone, F., Orezzi, P.G., Barbieri, W., Nicolella, V., and Penco, S. *Gazz. Chim. Ital. 97*, 1097, 1967.
7. Probst, G.W., Hoehn, M.M., and Woods, B.L. *Antimicrob. Agents. Chemother. 789*, 1965.
8. Takahashi, T., Sugawara, Y., and Susuki, M. *Tetrahedron Lett. 1873*, 1972.
9. Krowicki, K., and Lown, J.W. *J. Org. Chem. 52*, 3493, 1987.
10. Kissinger, K., Krowicki, K., Dabrowiak, J.C., and Lown, J.W. *Biochemistry 26*, 5590, 1987.
11. Lee, M.S., Pon, R.T., Krowicki, K., and Lown, J.W. *J. Biomol. Struct. Dyn. 5*, 939, 1988.
12. Lown, J.W. *Anti-Cancer Drug Design 3*, 25, 1988.
13. Lee, M., Shea, R.G., Hartley, J.A., Kissinger, K., Pon, R.T., Vesnaver, G., Breslauer, K.J., Dabrowiak, J.C., and Lown, J.W. *J. Am. Chem. Soc. 111*, 345, 1989.
14. Schultz, P.G., and Dervan, P.B. *J. Am. Chem. Soc. 105*, 7748, 1983.
15. Berman, H.M., Neidle, S., Zimmer, C., and Thrum, H. *Biochem. Biophys. Acta 561*, 124, 1979.
16. Gurskaya, G.V., Grokhovsky, S.L., Zhuze, A.L., and Gottikh, B.P. *Biochem. Biophys. Acta 563*, 336, 1979.
17. Remers, W.A., and Iyengar, B.S. In *Cancer Chemotherapeutic Agents*, Foye, W.O., (ed.) American Chemical Society, Washington, DC, 1995.
18. Zimmer, C., Kakiuchi, N., and Guschlbauer, W. *Nucl. Acids. Res. 10*, 1721, 1982.
19. Zimmer, C., Marck, C., and Guschlbauer, W. *FEBS Lett. 154*, 156, 1983.
20. Goodsel, D., and Dickerson, R.E. *J. Med. Chem. 29*, 727, 1986.
21. Zakrewska, K., Lavery, R., and Pullman, B. *J. Biomol. Struct. Dyn. 4*, 833, 1987.
22. Pullman, B. *Adv. Drug. Res. 18*, 1, 1989.
23. Mazurek, P., Feng, W., Shukla, K., Sapse, A.M., and Lown, J.B. *J. Biomol. Struc. Dyn. 9*, 299, 1991.
24. Martin, J.C., Wartell, R.M., and O'Shea, D.C. *Proc. Natl. Acad. Sci. USA*, 75, 5483, 1978.
25. Patel, D.J. *Eur. J. Biochem. 99*, 369, 1979.
26. Lown, J.W., Krowicki, B., Bhat, U.G., Skorogaty, A., Ward, B., and Dabrowiak, J.C. *Biochemistry 25*, 7408, 1986.
27. Bailly, C., Rommery, N., Houssin, R., and Hemchart, J.P. *J. Pharm. Sci. 78*, 910, 1989.
28. Sapse, A.M., Feng, W., Fugler-Domenico, L., Kabir, S., Joseph, T., and Lown, J.W. *J. Biomol. Struct. Dyn. 10*, 4, 709, 1993.

29. Kabir, S., and Sapse, A.M. *J. Comp. Chem. 12*, 1142, 1991.
30. Zimmerman, J., Rao, K.E., Joseph, T., Sapse, A.M., and Lown, J.W. *J. Biomol. Struct. Dyn. 9*, 599, 1991.
31. Spatola, A.F. *Chemistry and Biochemistry of Amino Acids, Peptides and Proteins*, *7*, 267, 1983.
32. Scheibye, S., Pederson, B.S., and Lawesson, S.O. *Bull. Soc. Chim. Belg. 87*, 229, 1978.
33. Bartlett, P.A., Spear, K.L., and Jacobsen, N.E. *Biochemistry 21*, 1608, 1982.
34. Kopka, M.L., Yoon, C., Goodsell, D., Pjura, P., and Dickerson, R.E. *Proc. Natl. Acad. Sci. USA 82*, 1376, 1985.
35. Coll, M., Frederick, C.A., Wang, A.H.J., and Rich, A. *Proc. Natl. Acad. Sci. USA 84*, 8385, 1987.
36. Sapse, A.M., Kabir, S., and Snyder, G. *THEOCHEM 339*, 227, 1995.

Appendix
Theoretical Studies of a Glucagon Fragment: Ser8-Asp9-Tyr10

ANNE-MARIE SAPSE[a,d], MIHALY MEZEI[b], DULI C. JAIN[c], and CECILLE UNSON[d]

Introduction

Glucagon is a polypeptide hormone involved in the production and degradation of glycogen by the liver. In insulin-deficient diabetes, blood glucagon levels are often abnormally high, implying that glucagon plays a role in the pathogenesis of diabetes.

Glucagon binding to its cell surface receptor is the first step in the events leading to the activation of adenylate cycles by the hormone. This binding is highly specific, as demonstrated by the lack of binding of related hormones secretin and vasoactive intestinal peptide. Moreover, small changes in the structure of the hormone lead to significant changes in the binding.

Unson et al. (1) have performed extensive studies on the effect of modifications of the primary structure of glucagon on its binding to the receptor and on its biological activity. They found that the aspartic acid residue Asp9 plays an important role in the signal transduction activity of glucagon, and when it is replaced by even a very similar residue such as glutamic acid, the potency decreases significantly. The replacement of Asp9 Glu does not greatly affect the binding to the receptor. Indeed, as shown by Unson et al. (2), the presence of a negative charge at the residue 9 ensures efficient binding to the receptor, presumably via a salt bridge formed of the carboxylate of glucagon with a positively charged group in the receptor. However, the transduction signal necessitates an aspartic residue at position 9, possibly as part of a His-Asp-Ser triad, which might create a charge relay network responsible for the signal (3). The simple existence of the carboxyl at this position, however, is not sufficient. Therefore,

[a] John Jay College and the Graduate School, City University of New York, 445 West 59th Street, New York, NY 10019.
[b] Mount Sinai School of Medicine, 1 Gustave Levy Place, New York, NY 10029.
[c] York College of the City University of New York, 94-20 Guy R. Brewer Boulevard, Jamaica, NY 11451.
[d] Rockefeller University, 1230 York Avenue, New York, NY 10031-6399.

150

it is of interest to estimate the difference in the positions of the carboxyl group of an aspartic residue from that of a glutamic residue.

The X-ray crystal structure of glucagon is known (4), but the conformation bound to the receptor is not clear as yet. Korn and Ottensmayer (5), largely based on Chou and Fasman probability calculations (6), have proposed a working model for glucagon in solution.

The similarity between aspartic and glutamic residues raises the question of possible larger differences in the preferential conformations of the two acids, large enough to affect a possible charge relay network. Sapse et al. (7) have performed ab initio quantum-chemical calculations to the study of both aspartic and glutamic acids. Geometry optimization has revealed that the two amino acids are in extended conformations. The superposition of the optimized structures of the two acids indicates that at the lipophilic CH_2 group position in the glutamic acid, the aspartic acid features the hydrophilic carboxylate ion. This difference might be responsible for the deactivation of glucagon upon replacement of Asp9 by a glutamic residue.

The present work studies the geometry of the tripeptides containing Asp9 and Glu9 in the middle, as fragments of glucagon and its Glu9 analogue. These are the Ser8-Asp9-Tyr10 tripeptide and the Ser8-Glu9-Tyr10 tripeptide. The tripeptides are investigated in the forms featuring both protonated and anionic aspartic or glutamic residues.

Protonated Form

Method and Results

The tripeptidic fragment of glucagon that contains the Asp9 residue in the middle was extracted via molecular modeling from the crystal structure of glucagon (8). The ends were truncated with one formyl group at one end and an amino group at the other. The fragment was geometry optimized with ab initio calculations at Hartree–Fock level, using the STO-3G minimal basis set, as implemented by the Gaussian-92 computer program (9). The aspartic residue was consequently replaced by a glutamic residue, and the geometry optimization was performed on the glutamic-containing tripeptide. The conformations so obtained are pictured in Figures A.1 and A.2, respectively. Their energies are given in Tables A.1 and A.2, respectively, and the optimized geometrical parameters are displayed in Table A.3. Figure A.3 depicts the superposition of the optimized conformations of the Ser-Asp-Tyr and Ser-Glu-Tyr. Figures A.4 and A.5 show the atomic numbering.

It was considered possible that the solvent interaction with the tripeptide could lead to the preferential stabilization of some conformations as opposed to others, and it was considered interesting to find out whether such an effect would be sufficient to make those conformations altogether

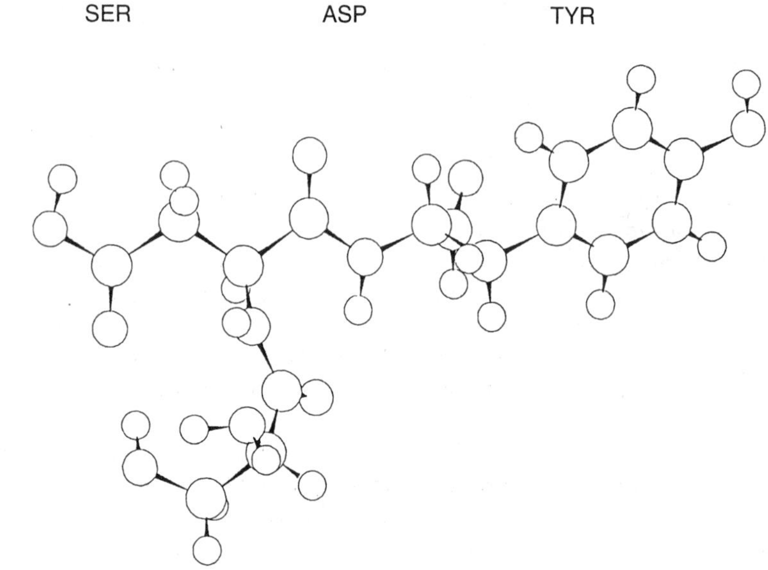

SER   ASP   TYR

FIGURE A.1 Protonated STO-3G optimized conformation of Ser-Asp-Tyr.

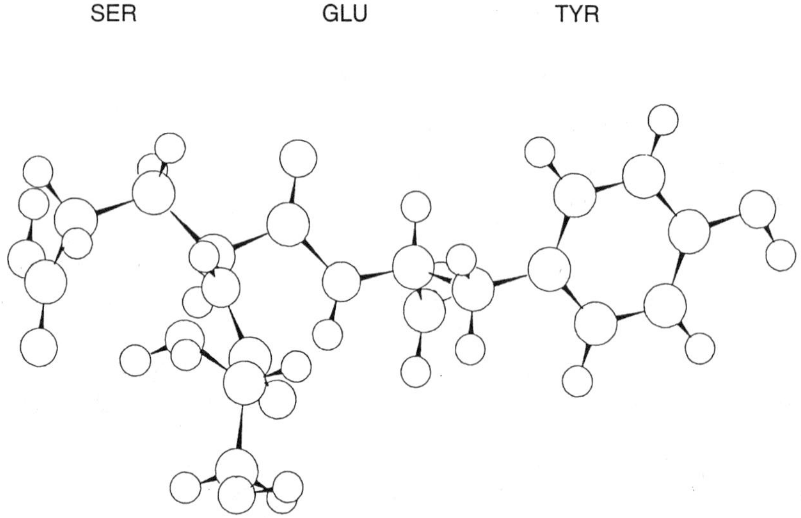

SER   GLU   TYR

FIGURE A.2 Protonated STO-3G optimized conformation of Ser-Glu-Tyr.

TABLE A.1 Calculated Energies for the Protonated Tripeptide Ser-Asp-Tyr Relative to the Optimized Geometry

Conformation		Ab Initio	Solvation
α^a	β^b	$E(STO\text{-}3G)^c$	$E(RF)^d$
168.6	241.2[e]	0.0[f]	0.0
0.0	0.0	18.5	−0.4
0.0	60.0	28.3	−0.8
0.0	180.0	18.8	−1.3
310.0	0.0	15.9	−0.3
310.0	60.0	25.3	−0.6
310.0	180.0	15.4	−1.1
268.0	0.0	14.2	−0.2
268.0	60.0	23.5	−0.0
268.0	180.0	13.3	−0.9
208.0	0.0	12.3	−0.1
208.0	60.0	21.3	−0.0
208.0	180.0	10.7	−0.0

[a] O1C2C1N1 angle (see Figure A.4).
[b] O3C5C4N2 angle (see Figure A.4).
[c] Hartree–Fock, STO-3G basis set (absolute energy: −1288.76803 au).
[d] Solvation energy estimate from reaction field (absolute energy: −0.3).
[e] Minimized energy conformation.
[f] All energies are in kcal/mole.

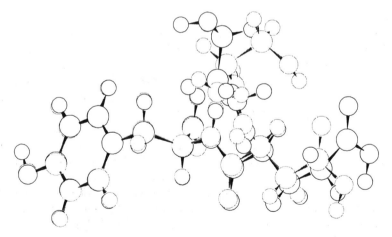

FIGURE A.3 Superposition of the optimized Ser-Asp-Tyr and Ser-Glu-Tyr (protonated form).

TABLE A.2 Calculated Energies for the Protonated Tripeptide Ser-Glu-Tyr Relative to the Optimized Geometry

Conformation		Ab Initio	Solvation
α^a	β^b	$E(STO\text{-}3G)^c$	$E(RF)^d$
168.6	241.2^e	0.0^f	0.0
0.0	0.0	14.1	−1.0
0.0	60.0	24.1	−2.0
0.0	180.0	14.3	−1.8
310.0	0.0	11.6	−0.8
310.0	60.0	21.7	−1.6
310.0	180.0	12.0	−1.7
268.0	0.0	10.3	−0.5
268.0	60.0	20.9	−1.3
268.0	180.0	10.5	−1.4
208.0	0.0	7.0	−0.4
208.0	60.0	16.6	−0.8
208.0	180.0	6.3	−1.2

^{a} O1C2C1N1 angle (See Figure A.5).
^{b} O3C5C4N2 angle (see Figure A.5).
^{c} Hartree–Fock, STO-3G basis set (absolute energy: −1327.34495 au).
^{d} Solvation energy estimate from reaction field (absolute energy: −0.3).
^{e} Minimized energy conformation.
^{f} All energies are in kcal/mole.

more stable. Accordingly, the solvent effect was estimated for a number of conformations that differ from each other by the values of the O1C2C1N1 angle (α) and the O3C5C4N2 angle (β). A grid method was used in which the angle α was given values of $0°, 310°, 268°$, and each of these values was used together with the values $0°, 6°$, and $180°$ for the angle β. Points where the energy was larger by more than 38 kcal/mole (0.06 au) were excluded from further consideration. The solute–solvent interaction energy was calculated with the method incorporated in the Gaussian-92 computer program, which implements the Onsager reaction field model. In this model, the solvent is considered a continuum characterized by its dielectric constant ε, and the solute occupies a spherical cavity of radius r. The electric field due to the solvent's dipole interacts with the solute's dipole. The Hamiltonian of the system contains a term due to solvation, which describes the coupling between the molecular dipole operator H and the reaction field R, which is a function of the dielectric constant and the cavity radius. The cavity's radius was determined by measuring the distance between the farthest atoms of the tripeptide, dividing it by two and adding to the van der Waals radius. Since the tripeptide is not exactly spherical, an error is thus introduced. However, since the error, which makes the cavity larger than it should be, is approximately the same for the different conforma-

TABLE A.3 Geometrical Parameters at 3-21G Calculational Level (Distances Are in au and Angles Are in Degrees)

Distances	Value	Angles	Value
		Ser-Asp-Tyr	
N1C1	1.491	C2C1N1	110.3
C1C2	1.559	02C3C1	111.1
C1C3	1.569	O1C2C1	123.5
C2O1	1.228	N2C2C1	113.7
C3O2	1.429	C4N2C2	122.4
C2N2	1.400	C5C4N2	112.7
N2C4	1.466	C6C4N2	110.3
C4C5	1.570	C7C6C4	111.5
C5O3	1.221	O4C7C6	124.9
C4C6	1.544	O5C7C6	116.0
C6C7	1.553	O3C5C4	124.0
C7O4	1.216	N3C5C4	111.1
C7O5	1.393	C8N3C5	122.4
C5N3	1.402	C9C8N3	110.1
C8N3	1.465	O6C9C8	124.2
C9C8	1.555	C10C8N3	111.2
C9O6	1.217	C11C10C8	112.8
C8C10	1.561	C12C11C10	121.1
C11C10	1.529	C13C11C10	121.1
C12C11	1.389	C14C13C12	120.1
C12C13	1.389	C15C14C13	119.4
C13C14	1.389	C16C15C14	120.1
C14C15	1.389	O7C14C13	117.6
C15C16	1.389	O1C2C1N1 (α)	158.6
C14O7	1.398	O2C2C1N1	38.7
		C4N2C2C1	197.3
		C6C4N2C2	149.0
		C5C4N2C2	273.6
		O3C5C4N2 (β)	241.2
		C7C6C4N2	298.1
		C9C8N3C5	243.3
		C10C8N3C5	119.4
		C12C11C10C8	257.6
N1C1	1.488	C2C1N1	110.1
C1C2	1.580	02C3C1	111.7
C1C3	1.560	O1C2C1	123.4
C2O1	1.227	N2C2C1	113.3
C3O2	1.435	C4N2C2	122.9
C2N2	1.399	C5C4N2	112.3
N2C4	1.469	C6C4N2	110.7
C4C5	1.571	C7C6C4	113.7
C5O3	1.220	C8C7C6	113.2
C4C6	1.548	C4C8C7	124.1
C6C7	1.548	O5C8C7	116.7
C7C8	1.552	O3C5C4	123.9
C8O4	1.215	N3C5C4	111.8
C8O6	1.397	C9N3C5	122.3
C5N3	1.404	C10C9N3	110.2

TABLE A.3 *Continued*

Distances	Value	Angles	Value
		Ser-Asp-Tyr	
C9N3	1.465	O6C10O9	124.3
C9C10	1.555	C11C9N3	111.1
C10O6	1.217	C12C11C9	112.8
C11C10	1.529	C13C12C11	121.2
C9C11	1.561	C14C12C11	120.5
C11C12	1.529	C15C13C12	120.2
C12C13	1.387	C16C14C12	121.3
C12C14	1.396	C17C15C13	120.0
C13C15	1.387	O7C17C15	123.2
C14C16	1.379	O1C2C1N1 (α)	203.0
C15C17	1.390	O2C2C1N1	300.3
C17O7	1.395	C4N2C2C1	185.8
		C6C4N2C2	154.8
		C5C4N2C2	280.6
		O3C5C4N2 (β)	243.3
		C10C9N3C5	244.5
		C11C9N3C5	121.7
		C12C11C9N3	184.4
		C13C12C11C9	260.7

tions, the conclusions are not likely to be altered. These results are also shown in Tables A.1 and A.2.

Discussion of Results

The lowest-energy conformation for Ser-Asp-Tyr features $\alpha = 158°$ and $\beta = 241.2°$. Among the single-point ab initio gas-phase calculated energies, some belong to conformations that are next to the most stable one. These conformations feature $\alpha = 208°$ and $\beta = 180°$.

For the glutamic-containing tripeptide, the most stable conformation features $\alpha = 203°$ and $\beta = 243°$. It can thus be seen that the β angle is practically the same for the aspartic- and the glutamic-containing tripeptides, while the α angle differs somewhat. Keeping the α angle to its optimum value with the β angle taking values of $0°$ and $180°$ increases the energy only by 6–7 kcal/mole. Also, for $\beta = 0°$ or $180°$ and α taking a value $268°$, the energy increases by 10 kcal/mole. For both tripeptides, $\beta = 60°$ is a particularly high-energy conformation, notwithstanding the value of α.

It can be seen from Figure A.3 that the tripeptide-containing aspartic residue differs by the Ser-Glu-Tyr tripeptide by the position of the carboxyl residue of aspartic or glutamic. This fact implies that even though this position is not essential for the binding, it is essential for activity.

Since these are neutral entities, the solvent effect of water is not as large as for the charged species. Using the reaction field method, the glutamic tripeptide features stronger solute–solvent interaction, no doubt because it

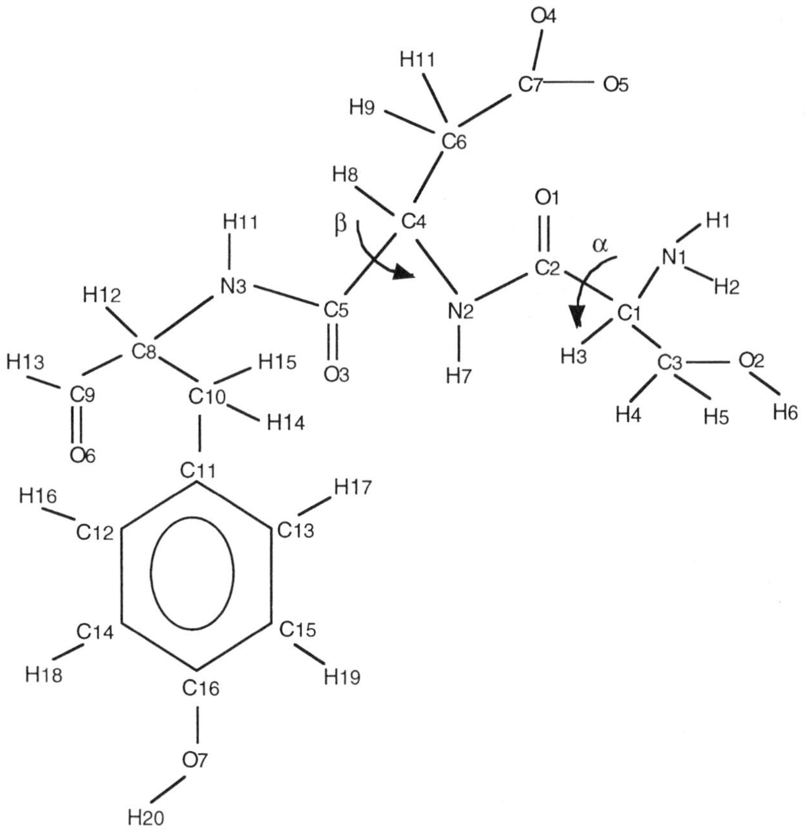

FIGURE A.4 Schematic form of Ser-Asp-Tyr with atomic numbering (anionic form).

has the polar carboxyl group closer to the wall of the solvent cavity. At no point can the differences in solute–solvent interaction energy in either approximation between different conformations supersede the differences in energy of the different conformations in gas phase. It may be concluded that the gas-phase optimized conformations are also the most stable in solution.

Anionic Species

Method and Results

The same fragment of glucagon in its anionic form obtained by removing the proton from the COOH group of the aspartic or glutamic residues has been subjected to Hartree–Fock geometric optimization using the STO-3G and the 3-21G basis sets. In addition, single-point calculations were per-

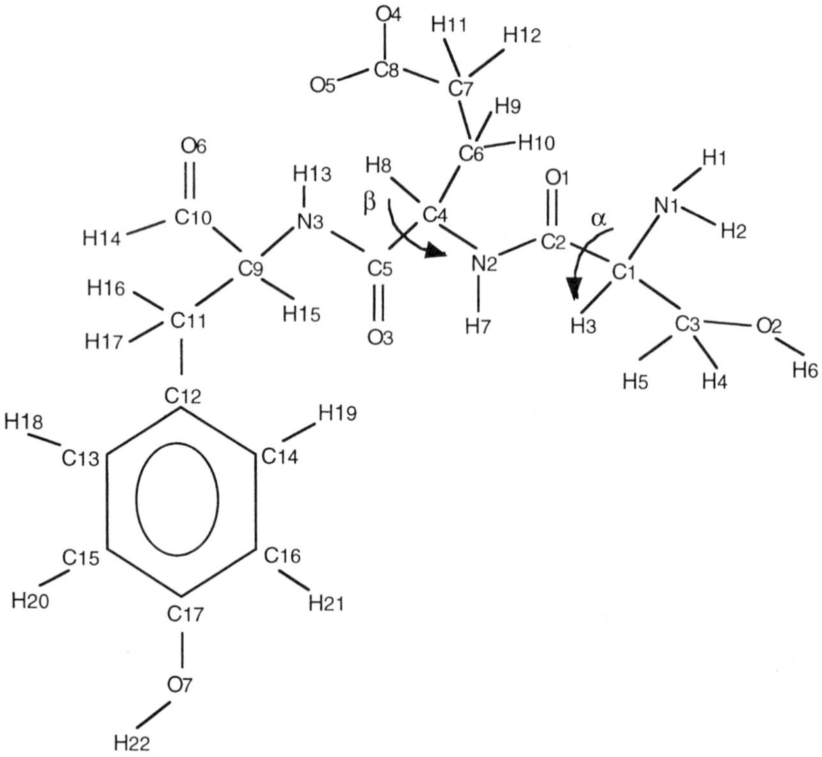

FIGURE A.5 Schematic form of Ser-Glu-Tyr with atomic numbering (anionic form).

formed for some of the most stable conformations at 6-31G* level and MP2/6-31G* level using the 3-21G optimized geometry. Figures A.6 and A.7 show the optimized conformations of the molecules. The minimum energies are given in Tables A.4 and A.5, respectively. The optimized parameters of both tripeptides are given in Table A.6.

The optimized structures of Ser-Asp-Tyr and Ser-Glu-Tyr were investigated in order to evaluate their interaction with water. The same methods were applied as for the protonated species, and the reaction field method also used the STO-3G basis set. In addition, the solvation energy was also estimated with the AMSOL-AMI (10, 11) program. AMSOL combines semiempirical quantum-chemistry calculations with continuum electrostatics and a surface-dependent dispersion, while AM1 gives the gas-phase semiempirical energy. We obtained the solvation energy as the difference between the AMSOL and AM1 energies. These results are also shown in Tables A.4 and A.5. Figure A.8 shows the superposition of the two anionic tripeptides in their most stable conformations. The energies of the investigated conformations at STO-3G, 3-21G, 6-31G*, and MP2/6-31G* levels, solvation energy estimates relative to the corresponding optimized geome-

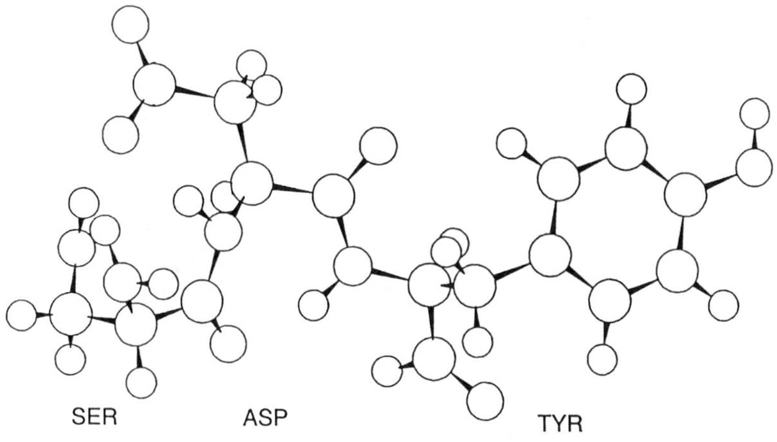

FIGURE A.6 Anionic 3-21G optimized conformation of Ser-Asp-Tyr.

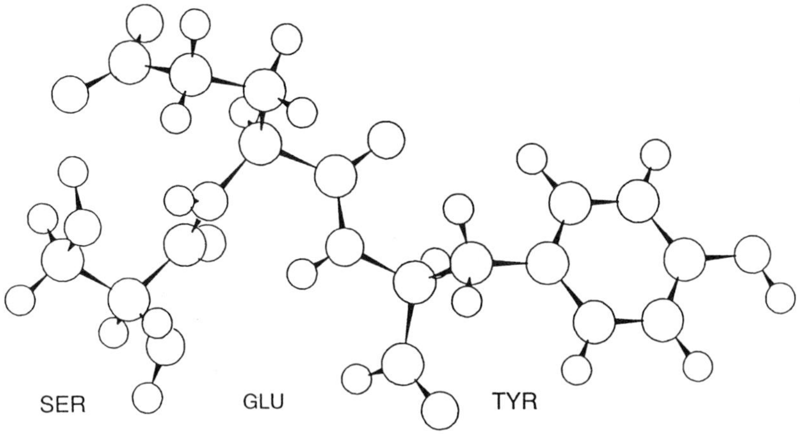

FIGURE A.7 Anionic 3-21G optimized conformation of Ser-Glu-Tyr.

tries, are shown in Tables A.4 and A.5 for the Ser-Asp-Tyr and Ser-Glu-Tyr, respectively.

Discussion of Results

It was found that for the anionic Ser-Asp-Tyr peptide the STO-3G basis set does not predict the correct geometry. Indeed, the proton from the N2 migrates to O5. When 3-21G calculations are performed, this does not take place. To verify that the migration is purely a basis set error, formamide was hydrogen-bonded to the formate ion, and it was found that by STO-3G geometry optimization, the proton migrates to the O of HCOO⁻. When the

TABLE A.4 Calculated Energies for the Anionic Tripeptide Ser-Asp-Tyr Relative to the Optimized Geometry

Conformation		Ab Initio				Solvation	
α^a	β^b	$E(STO\text{-}3G)^c$	$E(3\text{-}21G)^d$	$E(6\text{-}31G*)^e$	$E(MP2/6\text{-}31G*)^f$	$E(RF)^g$	$E(AMSOL)^h$
168.6	-122.9^i	0.0^j	0.0	0.0	0.0	0.0	0.0
161.3	0.0	10.1	15.6	17.6	16.4	-2.0	-2.8
161.3	90.0	15.1	22.0	14.7	18.1	-1.3	-2.2
161.3	180.0	6.9	7.8	11.7	12.2	-1.3	-1.1
0.0	-122.9	21.8	22.0	4.5	4.2	-1.7	-3.1
0.0	90.0	34.8	41.0	29.6	32.1	-0.6	-5.6

[a] O1C2C1N1 angle (see Figure A.4).
[b] O3C5C4N2 angle (see Figure A.4).
[c] Hartree–Fock, STO-3G basis set (absolute energy: –1288.0507 au).
[d] Hartree–Fock, 3-21G basis set (absolute energy: –1297.7339 au).
[e] Hartree–Fock, 6-31G* basis set (absolute energy: –1304.9880 au).
[f] Moller-Plesset/2, 6-31G* basis set (absolute energy: –1308.7732 au).
[g] Solvation energy estimate from reaction field (absolute energy: –5.27).
[h] Solvation energy estimate with AMSOL/AMI (10,11) (absolute energy: –67.5).
[i] Minimized energy conformation.
[j] All energies are in kcal/mole.

TABLE A.5 Calculated Energies for the Anionic Tripeptide Ser-Glu-Tyr Relative to the Optimized Geometry

Conformation		Ab Initio				Solvation	
α[a]	β[b]	E(STO-3G)[c]	E(3-21G)[d]	E(6-31G*)[e]	E(MP2/6-31G*)[f]	E(RF)[g]	E(AMSOL)[h]
115.8	168.2[i]	0.0[j]	0.0	0.0	0.0	0.0	0.0
115.8	0.0	5.8	12.1	23.2	22.0	1.3	-3.5
115.8	90.0	16.9	24.5	28.7	32.4	-0.3	-3.7
115.8	-90.0	1.7	4.5	27.9	33.0	-0.2	-0.4
180.0	0.0	41.6	41.2	10.8	11.9	-1.6	-13.6
180.0	-90.0	39.7	37.0	4.6	5.3	-2.2	-11.7
0.0	-90.0	39.5	40.8	25.7	28.6	-4.3	-11.9
180.0	168.2	40.0	36.7	24.6	27.1	-2.4	-13.0

[a] O1C2C1N1 angle (see Figure A.5).
[b] O3C5C4N2 angle (see Figure A.5).
[c] Hartree–Fock, STO-3G basis set (absolute energy: –1326.6139 au).
[d] Hartree–Fock, 3-21G basis set (absolute energy: –1336.5469 au).
[e] Hartree–Fock, 6-31G* basis set (absolute energy: –1344.0167 au).
[f] Moller–Plesset/2, 6-31G* basis set (absolute energy: –1347.9373 au).
[g] Solvation energy estimate from reaction field (absolute energy: –7.33).
[h] Solvation energy estimate with AMSOL/AM1 (10,11) (absolute energy: –64.65).
[i] Minimized energy conformation.
[j] All energies are in kcal/mole.

TABLE A.6 Geometrical Parameters at 3-21G Calculational Level (Anionic Forms) (Distances Are in au and Angles Are in Degrees)

Distances	Value	Angles	Value
		Ser-Asp-Tyr	
N1C1	1.461	C2C1N1	116.2
C1C2	1.529	C3C1N1	109.2
C2C3	1.552	O1C2C1	120.8
O1C2	1.236	O2C3C1	112.5
O2C3	1.427	N2C2C1	114.2
N2C2	1.341	C4N2C2	123.3
C4N2	1.474	C5C4N2	110.8
C5C4	1.528	C6C4C5	107.7
C6C4	1.535	O3C5C4	123.3
O3C5	1.227	C7C6C4	115.5
C7C6	1.557	N3C5C4	114.4
N3C5	1.349	O4C7C6	115.9
O4C7	1.298	O5C7C6	117.3
O5C7	1.223	C8N3C5	119.6
C8N3	1.449	C9C8N3	108.6
C9C8	1.511	C10C8N3	114.0
C10C8	1.553	O6C9C8	124.0
O6C9	1.208	C11C10C8	111.3
C11C10	1.514	O7C16C14	117.6
C12C11	1.387	O1C2C1N1 (α)	161.2
O7C16	1.384	O2C3C1N1	84.3
		N2C2C1O1	176.3
		C4N2C2C1	−143.8
		O3C5C4N2 (β)	−122.9
		N3C5C4O3	−178.7
		O4C7C6C4	40.0
		C8N3C5C4	−179.8
N1C1	1.487	C2C1N1	106.6
C1C2	1.517	C3C2C1	110.8
C2O1	1.220	O1C3C2	124.1
C1C3	1.541	O2C3C2	109.2
C3O2	1.436	N2C3C2	112.2
C2N2	1.354	C4N2C2	121.5
C4N2	1.448	C5C4N2	110.6
C5C4	1.519	C6C4N2	110.8
C6C4	1.554	N3C5C4	115.5
O3C5	1.226	C8C7C6	111.9
C7C6	1.541	C9N3C5	118.9
N3C5	1.355	O4C8C7	115.5
C9N3	1.451	O5C8C7	116.6
C8C7	1.550	C10C9N3	108.3
O4C8	1.279	C11C9N3	113.5
O5C8	1.233	O6C10C9	124.6
C10C9	1.510	O7C17C15	122.8
C11C10	1.541	O1C2C1N1 (α)	115.8
O6C10	1.211	N2C2C1O1	−177.1
C12C11	1.541	O2C3C1N1	47.2
C13C12	1.381	C4N2C2C1	−178.3
C14C12	1.393	O3C2C4N2 (β)	168.2
C15C13	1.387	N3C5C4O3	176.3
C16C14	1.393	O4C8C7C6	137.0
C17C15	1.379	C13C12C11C9	−99.5
O7C17	1.384		

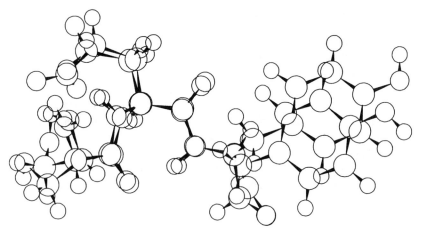

FIGURE A.8 Superposition of the optimized Ser-Asp-Tyr and Ser-Glu-Tyr (anionic form).

calculations were repeated with 3-21G and 6-31G* basis sets, the proton stayed on the N.

With geometry optimization using the 3-21G basis set, the Ser-Asp-Tyr tripeptide features an α angle of 161.25° and a β angle of −122.93°. Upon performing single-point calculations using a grid of values for α and β, it was found that only the points displayed in Table A.3 fall within 60 kcal/mole from the lowest-energy conformations. The other points considered, featuring α = 90°, are much higher in energy. The 6-31G* and MP2/6-31G* calculations at 3-21G optimized geometry show the same order of stability as the 3-21G//3-21G calculations. This justifies a posteriori our choice of smaller basis sets for the geometry optimization.

The Ser-Glu-Tyr tripeptide features the optimized α at 115.78° and the optimized β at 168.21°. However, a conformation with α at the optimized value of 115.78° and β at −90°, similar to the −122.93° found by optimization for the Ser-Asp-Tyr peptide, features an energy higher than that of the optimized conformation by only 4.4 kcal/mole.

When the reaction field method for the calculation of the solvent effect is applied, the optimized conformations for both tripeptides remain the most stable at all levels of ab initio theory used, as seen in Tables A.4 and A.5. However, for the AMSOL calculations, although the optimized conformation of Ser-Asp-Tyr remains the most stable, for the Ser-Glu-Tyr tripeptide the higher-level ab initio calculations predict α = 180° and β = −90° to be the most stable.

The major difference between the two methods is that the reaction field uses a sphere as the shape of the cavity, while the AMSOL method uses cavities closer to the shape of the molecules. Accordingly, the AMSOL method predicts similar interaction energies for such conformations as

$\alpha = 0$, $\beta = -90°$; $\alpha = 180°$, $\beta = -90°$; $\alpha = 180°$, $\beta = 0$; and $\alpha = 180°$, $\beta = 168.2°$. It is obvious that the value of β influences the AMSOL results less. The α values of 180° and 0° seem to afford a better contact with the solvent than the optimized 115.8° value. Since at higher calculational levels the structure with $\alpha = 180°$ and $\beta = -90°$ is quite stable, only 4-5 kcal/mole higher in energy than the optimized structure, the AMSOL-calculated solvent effect is sufficient to increase its stability in relation to the optimized structure. It becomes clear that the Ser-Asp-Try and the Ser-Glu-Tyr tripeptides are quite similar in their preferred backbone conformations, with $\alpha_{asp} = 161.3°$, $\alpha_{glu} = 180.0°$, $\beta_{asp} = -122.9°$, and $\beta_{glu} = -90.0°$. Figure A.8 shows the super-position of the most stable conformations in solution of the anionic Ser-Asp-Tyr and Ser-Glu-Tyr. It can be seen that the backbone superposes well, and the difference is the position of the carboxylate group of the Asp and Glu residues. Therefore, the difference in activity cannot be explained by their different backbone conformations in aqueous medium, but as observed for the acids themselves (7) by the fact that the negative carboxylate is set at a different position in Ser-Asp-Tyr than in Ser-Glu-Tyr. Even though this difference is too small to influence significantly the binding, it might be sufficient to cause a loss of adenylate cyclase activity.

References

1. Unson, C.G., Gurzanda, E.M., Iwasa, K., and Merrifield, R.B. *J. Biol. Chem. 264*, 789 and references within, 1989.
2. Unson, C.G., Macdonald, D., Ray, K., Durrah, T.L., and Merrifield, R.B. *J. Biol. Chem. 266*, 2762, 1991.
3. Unson, C.G., and Merrifield, R.B. *Proc. Natl. Acad. Sci. USA 91*, 454, 1994.
4. Sasaki, K., Dockerill, S., Adamiak, D.A., Tickle, I.J., and Blundell, T. *Nature 257*, 751, 1975.
5. Korn, A.P., and Ottensmeyer, F.P. *J. Theor. Biol. 105*, 403, 1983.
6. Chou, P.Y., and Fasman, G.D. *Biochemistry 14*, 2536, 1975.
7. Sapse, A.M., Mezei, M., Jain, D.C., and Unson, C.G. *J. Molecular Struct. (Theochem) 306*, 225, 1994.
8. Bernstein, F.C., Koetzle, T.F., Williams, G.J.B., Meyer, E.F. Jr., Brice, M.D., Rogers, J.R., Kennard, O., Shimanouchi, T., and Tasumi, M. The Protein Data Bank: A computer-based archival file for macromolecular structures. *J. Mol. Biol. 112*, 535–542, 1977.
9. Frisch, M.J., Trucks, G.W., Head-Gordon, M., Gill, P.M.W., Wong, M.W., Foresman, J.B., Johnson, B.J., Schlegel, H.B., Robb, M.A., Replogle, E.S., Gomperts, R., Andres, J.L., Raghavachari, K., Binkley, J.S., Gonzales, C., Martin, R.L., Fox, D.J., DeFrees, D.J., Baker, J., Stewart, J.J.P., and Pople, J.A. Gaussian 92. Gaussian Inc., Pittsburgh, PA, 1992.
10. Cramer, C.J., and Truhlar, D.J. *J. Am. Chem. Soc. 113*, 8305, 1991.
11. Cramer, C.J., and Truhlar, D. *J. AMSOL Program 606, QCPE*, Indiana University, Bloomington, Indiana, 1992.

Index

DATE DUE